T0293500

Conserving Biodiversity: Threats and Solutions (Volume II)

Conserving Biodiversity: Threats and Solutions (Volume II)

Devin Taylor

MURPHY & MOORE
www.murphy-moorepublishing.com

www.murphy-moorepublishing.com

ⓜMURPHY & MOORE

Cataloging-in-publication Data

Conserving biodiversity : threats and solutions (Volume II) / Devin Taylor.
    p. cm.
Includes bibliographical references and index.
ISBN 978-1-63987-729-4
1. Biodiversity conservation. 2. Biodiversity. 3. Ecosystem management. I. Taylor, Devin.
QH75 .C66 2023
333.951 6--dc23

Murphy & Moore Publishing
1 Rockefeller Plaza,
New York City,
NY 10020, USA

ISBN 978-1-63987-729-4

Contents

Preface

Biodiversity conservation refers to the protection, upliftment and management of biodiversity for ensuring sustainable benefits for the current and future generations. Biodiversity faces multiple threats at the global level such as pollution, urbanization, industrialization, depletion, erosion and increasing population. The loss of biodiversity severely affects the food chains of an ecosystem. There are three primary objectives of biodiversity conservation. These include protection and preservation of species diversity; sustainable utilization of species and resources; and prevention and restoration of ecological processes and life support systems. In-situ conservation and ex-situ conservation are the two methods of preserving biodiversity. The former involves the management and conservation of species in their natural habitats. Biosphere reserves, national parks, and wildlife sanctuaries are some protected areas for in-situ conservation. The conservation of endangered species in artificial ecosystems such as zoos, botanical gardens and gene banks is referred to as ex-situ conservation. The topics included in this book on conserving biodiversity are of utmost significance and bound to provide incredible insights to readers. The readers would gain knowledge that would broaden their perspective about this topic.

After months of intensive research and writing, this book is the end result of all who devoted their time and efforts in the initiation and progress of this book. It will surely be a source of reference in enhancing the required knowledge of the new developments in the area. During the course of developing this book, certain measures such as accuracy, authenticity and research focused analytical studies were given preference in order to produce a comprehensive book in the area of study.

This book would not have been possible without the efforts of the authors and the publisher. I extend my sincere thanks to them. Secondly, I express my gratitude to my family and well-wishers. And most importantly, I thank my students for constantly expressing their willingness and curiosity in enhancing their knowledge in the field, which encourages me to take up further research projects for the advancement of the area.

Devin Taylor

Biodiversity Estimates and the Importance of IGH

Eva Boon

Abstract In this chapter, I am concerned with the concept of Intra-individual Genetic Hetereogeneity (IGH) and its potential influence on biodiversity estimates. Definitions of biological individuality are often indirectly dependent on genetic sampling -and vice versa. Genetic sampling typically focuses on a particular locus or set of loci, found in the the the mitochondrial, chloroplast or nuclear genome. If ecological function or evolutionary individuality can be defined on the level of multiple divergent genomes, as I shall argue is the case in IGH, our current genetic sampling strategies and analytic approaches may miss out on relevant biodiversity. Now that more and more examples of IGH are available, it is becoming possible to investigate the positive and negative effects of IGH on the functioning and evolution of multicellular individuals more systematically. I consider some examples and argue that studying diversity through the lens of IGH facilitates thinking not in terms of units, but in terms of interactions between biological entities. This, in turn, enables a fresh take on the ecological and evolutionary significance of biological diversity.

10.1   Introduction to Intra-individual Genetic Heterogeneity

These days we have become beguiled with diversity: how animals, such as insects, and plants, such as angiosperms, have produced so incredibly many species. In the origins of multicellularity we see a most primitive example of diversification. In some ways, it is almost an ideal case because we can make an argument for its basis. (Bonner 1998)

Intra-individual genetic heterogeneity (IGH for short) is a characterisation that applies to multicellular biological entities. Simply put, it describes a state in which the cells of the biological entity under consideration contain divergent genomes. Some have argued that (similarity in) genome structure and content can give indications about the expected balance between cooperation and conflict between and

E. Boon (✉)
Eindhoven Technical University, Eindhoven, The Netherlands
e-mail: E.Boon@tue.nl

even within the cells (Queller and Strassmann 2009, 2016; Strassmann and Queller 2010) – in other words, about ecological interactions between genomes. This idea is a major rationale behind investigating biodiversity in the light of IGH.

Another rationale is that IGH highlights the fundamental elusiveness of some key concepts included in definitions of biodiversity. For example, the UN Convention on Biological Diversity defines biodiversity as "the variability among living organisms from all sources including, inter alia, terrestrial, marine and other aquatic ecosystems and the ecological complexes of which they are part; this includes diversity within species, between species and of ecosystems".[1] The introduction to this volume already mentioned that the ambiguity surrounding terms such as "organisms", "species" and "ecosystems" has already been discussed for a longer time. Technological advances now enable philosophers of biology and biologists to consider in detail how actual patterns of genetic diversity match or clash with these concepts. Studying IGH is part of this effort, in the sense that patterns of genetic diversity within and between biological entities do not always coincide with our perception of these entities as biological, ecological or evolutionary units. To stay with the analogy of diagnosing and treating a patient, as was elaborated in the introduction, understanding IGH is important to better diagnose a patient. The consequences for treatment will be discussed at the end of this chapter.

Here, I am concerned with how the concept of IGH can influence our perception of biodiversity, considering that in biodiversity studies, our definition of biological individuality is often dependent on genetic sampling -and vice versa. Genetic sampling is often concentrated on a particular genome, such as the mitochondrial, chloroplast of nucelar genomes. However, if ecological function or evolutionary individuality can play out on the level of multiple divergent genomes, as I shall argue is the case in IGH, our current genetic sampling strategies may miss out on relevant biodiversity.

I will proceed as follows. As more and more examples of IGH become available, it is becoming possible to investigate the positive and negative effects of IGH on the functioning of a multicellular individual (sect. 10.2). I argue that considering diversity through IGH facilitates thinking not in terms of units, but in terms of interactions between biological entities (sect. 10.3). This, in turn, may enable a fresh take on the ecological and evolutionary significance of diversity. From the examples proposed in this chapter, we can consider how IGH as an unexplored dimension of biodiversity may help us understand what diversity is *relevant* to our research goals.

10.2   Examples of IGH

To make the concept of IGH more concrete, I will start off with a number of familiar and possibly less familiar examples of IGH. Common resolutions to genetic conflict within the organism, such as the separation of germline and soma and genetic

---

[1] https://www.cbd.int/doc/legal/cbd-en.pdf

bottlenecks, are only applicable to a narrow range of multicellular organisms -mostly metazoans. Organisms that consist of easily regenerating parts (plants, algae) or hyphal networks (fungi) seem often more genetically heterogeneous. In this context, it is interesting to note that metazoans that regenerate from body fragments (medusa, sponges, urochordates) are also more often reported to be genetically heterogeneous (Rinkevich 2004, 2005).

Cases of intra-individual genetic heterogeneity can be divided in chimeras and mosaicisms (Pineda-Krch and Lehtila 2004a; Santelices 2004a), of which examples will be discussed separately below. In the case of mosaic entities, genomes are divergent but homologous in the sense that they share a recent common ancestor. In the case of chimeras, the genetic heterogeneity is nonhomologous: cells may have originated from evolutionary highly divergent lineages. From this definition, it becomes apparent that the distinction between a mosaic and a chimeric biological entitity is ultimately a judgment on evolutionary divergence, which in itself is often based on genome similarity.

10.2.1 Mosaic Individuals

What kinds of individuals are mosaic? As mentioned above, genetic differentiation between somatic cells is often found within plants, animals and fungi that propagate by cloning of body parts. The prevalence of mosaicism in clonally reproducing plants and animals is easy to explain if mosaicism correlates with mutation rate and longevity (Gill et al. 1995). Of course, the mosaic state will be cut short when a single cell bottleneck occurs in the reproductive cycle. Still, this does not seem to stop mosaicism from occurring in multicellular entites that pass through a single cell bottleneck, i.e. metazoans such as fish (Matos et al. 2011) and corals (Schweinsberg et al. 2015).

Mosaicism in humans is a burgeoning field, since much of present-day cancer research relies on assessing genetic heterogeneity between tumor cells, which in turn determines (in some cancers) much of the treatment and prognosis. This approach relies on the argument that the genetic heterogeneity in the tumor is governed by different evolutionary dynamics as the rest of the body (see for example Jacobs et al. 2012; Laurie et al. 2012; Vijg 2014 and references therein). Mosaicism can also have much less dramatic influences in humans, and increasing reports on mosaicism in humans (Youssoufian and Pyeritz 2002; Erickson 2014; Spinner and Conlin 2014) and even the germline (Samuels and Friedman 2015; Jónsson et al. 2017) seem to underline the varying evolutionary outcomes for IGH in a multicellular individual: positive, negative or neutral.

Another example demonstrates how mosaicism can be deeply integrated in the life history of a biological entity. Arbuscular mycorrhizal fungi (AMF) are an ancient phylum of heterotrophs that form symbioses with the majority of land plants (Wang and Qiu 2006). AMF were reported to contain hundreds or thousands of genetically differentiated nuclei within the same cytoplasm (Kuhn et al. 2001; Hijri and Sanders 2005; Boon et al. 2010, 2013b, 2015). The exact number of genetically differentiated nuclei is not clear, since genetic variation has never been exhaustively sampled for any locus in these fungi. No sexual stage has been observed, although possible recombination has been reported in AMF (see Riley et al. 2016 and references therein).[2]

It is possible that a single nucleus is not a viable entity in AMF –only populations of nuclei are (see Boon et al. 2015; Wyss et al. 2016 and references therein). For example, spores of *Rhizophagus irregularis* do not germinate under a certain volume, which is positively correlated with the number of nuclei within the spore (Marleau et al. 2011). Some authors have proposed that genetic differentiation between nuclei within the AMF cytoplasm is maintained through the fusion of related hyphae, or anastomosis (Giovannetti et al. 2015 and references therein), and is lost at sporulation (Boon et al. 2013b). This means that AMF are both mosaic individuals in the sense that their nuclei share the same genealogy, and chimeras in the sense that at least some of this variation is the result of hyphal fusion between related hyphal systems.

The positive correlation between anastomosis rates and level of relatedness between hyphae supports the view that AMF do form genetically delineable entities –although maybe not on a genome level, but on that of the genome population or pangenome. This has also been suggested by Boon et al. (2015). The propensity to fuse within cultures of the same strain seems diminished by drift (Cárdenas-Flores et al. 2010). This may indicate that the nuclei within an AMF hyphal system show self-nonself recognition *as a population*. Finally the composition of the genome population has an influence on the genotype: a change in nuclear population through anastomosis changes the (symbiotic) properties of a strain (Sanders and Rodriguez 2016 and references therein).

To summarise this example, the AMF phenotype seems determined by not a single, but multiple coexisting genomes. Anastomosis, or a lack thereof, can change the AMF phenotype, which is in turn selected upon by its environment (Roger et al. 2013b, a; Angelard et al. 2014; Wyss et al. 2016; Sanders and Rodriguez 2016). AMF form symbioses with almost all land plants (Wang and Qiu 2006), and the age of their evolutionary lineage (an estimated 500 million years, coinciding with the rise of land plants (Corradi and Bonfante 2012)) testifies to the potential ecological impact and longevity of IGH as an evolutionary strategy. The presence of mycorrhiza in the soil confers a inestimable fitness advantage to plant communities: most plant taxa form symbioses with AMF in which posphorus is exchanged for plant-produced sugars, thus stimulating plant growth and overall community biodiversity (van der Heijden et al. 2016)

In this example on AMF, IGH seems to be an essential to understand AMF life history, ecology and evolution. At the same time, the precise effects of IGH on AMF life history are hard to estimate since this IGH is often not taken into account in experimental setups and field studies due to technical and conceptual challenges (Sanders and Rodriguez 2016). Still, we can add yet another layer of complexity. If

---

[2] Note that some recombination estimates may not be reliable if Glomus is indeed multigenomic.

we consider that AMF are obligate plant symbionts and themselves are associated with particular microbial communities, a picture emerges of another set of interactions. AMF are functionally dependent on their plant hosts, and probably gain major fitness benefits from their associated microbial communities (Bonfante and Anca 2009; Herman et al. 2011). Thus, since the ecological function and evolutionary longevity of the mosaic AMF is dependent on nonhomologous lineages (i.e. plants and microbes), we may also consider AMF and their associated plants and microbes as chimeras. This is not only theoretically relevant: the growth benefits that AMF confer to their plant hosts can potentially confer huge benefits to sustainable agriculture.

10.2.2 *Chimeric Individuals*

As in the AMF example above, most mutualisms can also be considered chimeras. This depends on the criteria for individuality that are being used, and to some extent on the degree of genetic divergence one is willing to accept between the component genomes of the chimera. For example, in the case of lichen and corals the mutualism is so tight that the historical and most intuitive view is to see the chimera as a single entity. Only with the advent of molecular techniques have scientists started to distinguish different genomes and consider the partners as separate 'individuals'. With the following two examples, I would like to illustrate how broad the definition of 'chimera' can be, and highlight how considering IGH has consequences for ecological and evolutionary inference in these examples.

My first case is chimerism in macroalgae, which is a nonmonophyletic group that encompasses brown, green and red algae (Santelices 2004b). Here, I will focus on red algae or Rhodophyta, since IGH in this taxon has been most extensively documented. The algae germinate from a disk, which can originate from multiple spores. These spores may fuse, or form individual cell walls that are subsequently surrounded by a thickened common wall. This process, called coalescence, occurs often in red algae (Santelices et al. 1999) but not between different species (Santelices et al. 2003). Coalescing disks, or crusts, may or may not fuse with each other or with new algal sporelings. Various fitness advantages were demonstrated for coalescencing disks compared to unitary disks (which have originated from a single spore). Fusion decreases the probability of mortality in early growth (Santelices and Alvarado 2008), improves erect axis formation and growth (Santelices et al. 2010) and confers an advantage later in the life cycle through differences in branching and fertility (Santelices et al. 2011). Coalescence and a fitness advantage for coalescing disks have also been reported, although less extensively, in brown (Wernberg 2005) and green (Gonzalez and Santelices 2008) algae.

Thus, like in AMF, we encounter a population of genotypes (from multiple haploid spores) which together creates a polyploid phenotype that is selected upon as a single entity (Monro and Poore 2004). In red algae, these polyploids break up again at sporulation. Possibly, selection (for cooperation) between haploid genomes occurs at the formation of the disk. How this selection takes place is unclear. It seems reasonable to suppose that selection occurs for compatibility between particular loci or even entire genomes. It is important to note that, like in the previous AMF example, it is not possible to associate the phenotype of a red alga with a particular unchanging set of spore haplotypes. The phenotype of interest, i..e the disk and the thalli that grow from it, is based on a *varying* population of multiple haploid nuclear genomes. Therefore, the phenotype of the macroalga cannot be reduced to its component genomes. It is in the *interaction* between these varying genomes that a unique phenotype is established.

The ecological and evolutionary consequences of the chimeric state in macroalgae are not easy to disentangle, even more so because seaweeds can reproduce clonally as well (Fagerström and Poore 2000; Collado-Vides 2001). Nevertheless, the above description makes clear that the life history and evolutionary constraints of coalescing red algae cannot be described accurately without taking chimerism into account. Again, in the light of the ecological and agricultural importance red algae this is not a merely academic preoccupation. For example, multisporic coalescing recruits have higher survival rates (Santelices and Aedo 2006). Thus, IGH as a state can be manipulated by farmers to increase higher yields.

My second example of chimerism is the case of microbial multi-species consortia or communities. With the advance of molecular techniques it has become possible to study microbial communities in more detail than ever before, even though passing from the fase of amassing vast quantities of data to that of interpreting them has proven to be a challenge. Here, I would like to highlight a few patterns that have emerged with respect to microbial diversity and function and that are relevant to the concept of IGH. The following three points are discussed in more detail in Boon et al. (2013a).

First, microbial taxa are hard to circumscribe precisely, for a number of well-described ontological and epistemological reasons.[3] For the purposes of this chapter, it suffices to state that microbial taxonomy is heavily dependent on the molecular biology toolbox. This toolbox, although indispensible, has a number of limitations. The most relevant limitation for my point is that especially early conceptions of microbial taxa heavily relied on the assumption of genome stability. And there's the rub: in many microorganisms, genomes can change rapidly through gene loss, gene duplication, and the acquisition of genes from distant lineages via lateral gene transfer (LGT).

A second pattern is that taxonomic or phylogenetic thresholds (e.g. 3% genetic differentiation) for taxon delineation fail to adequately delineate ecologically cohesive units. Even though a unifying species concept is not strictly needed for ecological analysis, also a pluralist stance needs a sound rationale and consistent approach (or set of approaches) to define 'units'. Unfortunately, microbial diversity and community function do not always correlate. Microbes rarely act alone and are often interdependent. It is possible that less than 1% of all known microbes can be successfully cultured on their own (Staley and Konopka 1985), an observation also known as 'Great Plate Count Anomaly'. It is now clear that many microbes depend on the activity of other microbes to successfully grow and reproduce via mecha-

---

[3] Doolittle (2013) has written an extensive review on the history and challenges of microbial ontology and of course O'Malley (2014) is an invaluable resource here as well.

nisms including acquisition and exchange of metabolites (references in Boon et al. 2013a).

A final tendency is that microbial function may be a property of communities as well as of cells. Particular metabolic capacities might not be encoded within a single microbial cell, or even within a single type of microbial cell. Instead, there is increasing evidence that many microbial functions are encoded by gene networks in which genes may be easily replaced by functionally equivalent but phylogentically distant alternatives. These gene networks may be found in varying sets of microbial taxa, without a single taxa being characterised by a particular set of genes or functions. We face the same situation as in the previous examples, in which no single community genotype codes for a single community phenotype.

If microbial communities can be understood as 'chimeras', it might not be possible to lead a community function (for example, a particular metabolic product or process) back to a single taxonomic group (but see Inkpen et al. 2017). The diversity of microbes is now being explored using surveys that draw on hundreds or thousands of samples and controlled experiments, with rapid genetic assessment techniques providing much of the evidence for taxonomic and functional diversity. Since microbial interactions span all taxonomic ranks, from strain to superkingdom, understanding microbial diversity then seems to necessitate a community-centric approach (Zarraonaindia et al. 2013).

Mechanisms for the evolution of interdependence within microbial communities have been proposed in the form of a Public Goods Hypothesis ((McInerney et al. 2011) and the Black Queen Hypothesis (Morris et al. 2012). The authors of the ITSNTS model ('It's the song, not the singer') even propose 'casting metabolic and developmental interaction patterns, rather than the taxa responsible for them, as units of selection' (Doolittle and Booth 2016): in other words, microbial interaction patterns are stabler units of selection than the microbial cells that produce these patterns. For a more in-depth discussion of the evolutionary and ecological implications of seeing microbial communities such as biofilms as evolutionary individuals, see Boon (in preparation).

10.2.3 Mosaic vs. Chimeric Individuals

The reader might wonder by now whether the distinction between mosaic and chimeric individuals is at all relevant. Is the difference between the two not just a matter of degree of relatedness between individuals, rather than a difference in kind of individual? The answer to this ontological question is not at all straightforward. However, from an epistemic point of view, the differences between these two types of intraindividual genetic heterogeneity are relevant to the practise of evolutionary inference.

For example, fitness calculations between mosaic entities and chimeras are performed differently. If genetically differentiated but related cell lines work together, as in mosaic individuals, a case could be made for a special sort of kin selection. After all, there is a considerable chance that gene variants between related lineages are shared. However, if unrelated cell lines become integrated in a single entity, the balance of costs and benefits that ultimately decides between competition and cooperation cannot be explained by a more than average chance to transfer one's own genes as present in the other.

Some might disagree that there is even an epistemic difference between chimeras and mosaic biological entites. Multilevel selection theory (see Okasha 2006 and references therein) stresses that kin selection is really just a special case of group selection. Interestingly, this discussion is also highly relevant within microbial community ecology and evolution, in which the question whether microbial communities can evolve is a hot topic of debate. Boiled down to its essence, this question is really about whether entities that are composed of nonhomologous lineages can evolve as a single unit–and whether is this is a useful question to ask (Boon in preparation).

10.3   The Importance of IGH in Ecology and Evolution

Above examples lead me to two main themes for the relevance of IGH in biodiversity research. The first is that multiple *varying* genotypes can lead to a single phenotype. Since the phenotype is the actual set of traits that is selected upon, or that is ecologically relevant, extreme caution should then be exercised when a single genome (or genotype, or even a simple barcode) is taken as a proxy for the phenotype. If the more complicated genotype-phenotype relationship that is implied by IGH is ignored and genotypes that are associated with a particular phenotype are inadequately sampled, it will be difficult if not impossible to find reproducible patterns and predict community composition or ecosystem function. Second, if one of the aims of measuring biodiversity is to predict or at least understand ecosystem function, it is vital to note that while community ecology considers interactions among entities, the inference of these interactions depends critically on the level at which entities are defined.

These two themes may be made more concrete with an example: while it may be possible to describe a microbial community as performing a single ecosystem function, it may not be possible to find a specific genotype or even set of genotypes stable enough (i.e. reoccurring consistently) to characterise this functional unit. Instead, one may want to consider whether instead particular interactions between units, such as a particular exchange of metabolites or another shared fitness benefit, may be the most stable component of the interaction.[4]

This situation is a radical departure from a more traditional view, in which a one-to-one relation is assumed between genotype and phenotype. In other words, once we look away from our metazoan bias it may no longer be possible to explain phenotype and its ecological role by measuring the genotype, since this genotype, even as an amalgam of multiple component genotypes, is simply not stable enough.

---

[4] See also the ITSNTS argument of Doolittle and Booth (2016).

10.3.1 The Metazoan Bias

The examples in the previous section might seem "atypical" in the context of biodiversity. In fact, when speaking of biodiversity there is often a bias towards species that are relatively easy to identify and delineate, such as animal species. Yet vast diversity, however measured, can be found in groups such as algae, fungi, and the many phylae of microbes and virusses, which are often at the basis of ecosystem function (e.g. Wagg et al. 2014).

Still, one should wonder whether it makes sense to describe above examples as instances of IGH. In other words, how permissive can a definition of biological or evolutionary individuality be without losing its use? The term 'Intra-individual genetic heterogeneity' ultimately pivots on the definition of the 'individual'. To determine an ecological function or identify an evolutionary process, one needs to distinguish the entities that perform these functions or processes.

The discussion on biological delineation and individuation has been conducted in different contexts already and has taken a central place in recent philosophy of biology discussions (see for example Queller and Strassmann 2016; Clarke 2016; Pradeu 2016 and references therein). It becomes clear from these recent considerations that there are valid reasons to consider biological organisation from many different viewpoints. In other words, different research goals justify the use of divergent concepts of biological or evolutionary individuality and thus warrant a pluralist approach.

In this context, it does make sense to describe different kinds of biological entities as instances of IGH. For example, when we consider a system with AMF, we could choose to look at a single AMF nucleus, at a population of nuclei, or at an entire hyphal system. Enlarging our scope even more, we could choose to include the plant partners as well as the surrounding microbial communities. I propose that it is in this choice that the real point of discussion lies: how to decide on the relevant unit of diversity?

10.3.2 Biological Organization, Hierarchy and Relevance

A genotype, or even the entire genome, is often for practical purposes employed as a unique identifier for 'the biological individual'. Of course, this biological individual cannot be fully described by only its genetic code. If this were so, we would consider human identical twins to be one and the same biological individual. However, although the individual is not defined by a unique genome, a unique genome seems to havebeen a convincing criterion for assigning individuality. Why?

One possible reason is that the organization of biological diversity is considered to be hierarchical. In this view, DNA is organized in cells, cells in bodies, bodies in populations and populations in species. It is implied that without cells competing or cooperating, the body would not exist, and without bodies competing or cooperating, a population would not exist, and so forth. Leo Buss, for example, stated that "[An] explicitly hierarchical perspective on evolution predicts that the myriad complexities of ontogeny, cell biology and molecular genetics are ultimately penetrable in the context of an interplay of synergisms and conflicts between different units of selection" (Buss 1987).

This idea of hierarchy is also prominent in the literature on major transitions in evolution. Maynard Smith and Szathmáry proposed that complexity in evolution increases with time, which is achieved through a series of major transitions. They also described this complexity as mostly hierarchical (Maynard Smith and Szathmáry 1995). Others have continued or varied on this view of evolution of life on earth, yet all agree that cooperation and competition takes place at definable 'levels' (Clarke 2016 and references therein). A formalisation of this view can be found in multilevel selection theory. Proponents of this theory aim to develop and formalize the tools we need to describe and quantify the relative importance of different levels of selection (e.g. Wilson and Sober 1994). Ultimately, the interactions between these levels are proposed to lead to the diversity we observe among biological entities.

Much discussion in the major transitions literature is then about finding out how conflict at a particular level of organization is resolved, in order to explain the evolution and diversity of another level of biological organization. In this manner, a hierarchical view on biological diversity can offer a perspective with the scope to explain a large number of observations. However, it can also lead to misleading assumptions or obscure similarities. For example, the link of genome homogeneity with the delineation of the biological individual is based on the assumption that IGH leads to conflict within that individual (Michod and Roze 2001; Strassmann and Queller 2004). Yet it is not clear whether genome heterogeneity always leads to conflict. It is possible that there are cases where IGH can actually confer an advantage to the multicellular community it is part of. Some of these examples were already discussed above. The question then becomes more nuanced: when is IGH relevant?

The simple answer may be: when there is a significant effect of IGH on the possible evolutionary trajectories (sometimes referred to as 'evolvability') and ecological range that a biological unit can follow or occupy as a result of its IGH. These latter two concepts are exactly what is at stake in many biodiversity investigations. Moreover, some of the ways in which biodiversity is understood are based on taxonomic or ecologic hierarchies (e.g. Sarkar 2002). Red algae and arbuscular mycorrhizal fungi are two fairly well-documented organisms in which intra-individual heterogeneity plays an important role in understanding of both evolvability and ecological range. Furthermore, even though IGH is sparsely documented, reviews are available with more examples (Santelices 1999; Pineda-Krch and Lehtila 2004b, a), as well as a range of suggestions on how IGH could affect life history (Pineda-Krch and Lehtila 2004b; Folse 2011; Folse and Roughgarden 2012). Finally, the importance and relevance of IGH should be decided on a case-by-case basis –without assuming or dismissing its potential role off-hand.

10.4   Conclusions

I argued that the shortcut one genome-one individual has closed our eyes to the possible importance of IGH in evolution and evolution –and thus to its role in for biodiversity estimates. Arguing from the examples in this chapter, I propose that IGH can help us understand what diversity is *relevant* to our research goals. To maintain the analogy from the introduction: which characteristics of the patient and her symptoms are relevant to a diagnosis and treatment?

By expanding our practical and conceptual tools to facilitate the study of genetic heterogeneity at many different levels of biological organisation, we can start to understand diversity by focusing on interactions between entities –however defined.

References

Angelard, C., Tanner, C. J., Fontanillas, P., et al. (2014). Rapid genotypic change and plasticity in arbuscular mycorrhizal fungi is caused by a host shift and enhanced by segregation. *The ISME Journal, 8,* 284–294.

Bonfante, P., & Anca, I.-A. (2009). Plants, mycorrhizal fungi, and bacteria: A network of interactions. *Annual Review of Microbiology, 63,* 363–383. https://doi.org/10.1146/annurev.micro.091208.073504.

Bonner, J. T. (1998). The origins of multicellularity. *Integrative Biology Issues News and Reviews, 1,* 27–36. https://doi.org/10.1002/(SICI)1520-6602(1998)1:1<27::AID-INBI4>3.0.CO;2-6.

Boon, E. (in preparation). *Biofilms as evolutionary individuals: An empirical question?*

Boon, E., Zimmerman, E., Lang, B. F., & Hijri, M. (2010). Intra-isolate genome variation in arbuscular mycorrhizal fungi persists in the transcriptome. *Journal of Evolutionary Biology, 23,* 1519–1527. https://doi.org/10.1111/j.1420-9101.2010.02019.x.

Boon, E., Meehan, C. J., Whidden, C., et al. (2013a). Interactions in the microbiome: Communities of organisms and communities of genes. *FEMS Microbiology Reviews, 38,* 90–118. https://doi.org/10.1111/1574-6976.12035.

Boon, E., Zimmerman, E., St-Arnaud, M., & Hijri, M. (2013b). Allelic differences within and among sister spores of the arbuscular mycorrhizal fungus Glomus etunicatum suggest segregation at sporulation. *PLoS One, 8,* e83301.

Boon, E., Halary, S., Bapteste, E., & Hijri, M. (2015). Studying genome heterogeneity within the arbuscular mycorrhizal fungal cytoplasm. *Genome Biology and Evolution, 7,* 505–521.

Buss, L. W. (1987). *The evolution of individuality.* Princeton: Princeton University Press.

Cárdenas-Flores, A., Draye, X., Bivort, C., et al. (2010). Impact of multispores in vitro subcultivation of Glomus sp. MUCL 43194 (DAOM 197198) on vegetative compatibility and genetic diversity detected by AFLP. *Mycorrhiza, 20,* 415–425.

Clarke, E. (2016). A levels-of-selection approach to evolutionary individuality. *Biology and Philosophy, 31,* 893–911. https://doi.org/10.1007/s10539-016-9540-4.

Collado-Vides, L. (2001). Clonal architecture in marine macroalgae: Ecological and evolutionary perspectives. *Evolutionary Ecology, 15,* 531–545. https://doi.org/10.1023/a:1016009620560.

Corradi, N., & Bonfante, P. (2012). The arbuscular mycorrhizal symbiosis: Origin and evolution of a beneficial plant infection. *PLoS Pathogens, 8,* e1002600. https://doi.org/10.1371/journal.ppat.1002600.

Doolittle, W. F. (2013). Microbial neopleomorphism. *Biology and Philosophy, 28,* 351–378. https://doi.org/10.1007/s10539-012-9358-7.

Doolittle, W. F., & Booth, A. (2016). It's the song, not the singer: An exploration of holobiosis and evolutionary theory. *Biology and Philosophy, 32*, 5–24. https://doi.org/10.1007/s10539-016-9542-2.

Erickson, R. P. (2014). Recent advances in the study of somatic mosaicism and diseases other than cancer. *Current Opinion in Genetics & Development, 26*, 73–78. https://doi.org/10.1016/j.gde.2014.06.001.

Fagerström, T., & Poore, A. G. B. (2000). Intraclonal variation in macroalgae: Causes and evolutionary consequences. *Sel, 1*(1), 123–133.

Folse, H. J. (2011). *Evolution and individuality: Beyond the genetically homogenous organism.* Stanford: Stanford University.

Folse, H. J., & Roughgarden, J. (2012). Direct benefits of genetic mosaicism and intraorganismal selection: Modeling coevolution between a long-lived tree and a short-lived herbivore. *Evolution, 66*, 1091–1113. https://doi.org/10.1111/j.1558-5646.2011.01500.x.

Gill, D. E., Chao, L., Perkins, S. L., & Wolf, J. B. (1995). Genetic mosaicism in plants and clonal animals. *Annual Review of Ecology and Systematics, 26*, 423–444.

Giovannetti, M., Avio, L., & Sbrana, C. (2015). Functional significance of anastomosis in arbuscular mycorrhizal networks. In T. R. Horton (Ed.), *Mycorrhizal networks* (pp. 41–67). Dordrecht: Springer.

Gonzalez, A. V., & Santelices, B. (2008). Coalescence and chimerism in Codium (Chlorophyta) from central Chile. *Phycologia, 47*, 468–476. https://doi.org/10.2216/07-86.1.

Herman, D. J., Firestone, M. K., Nuccio, E., & Hodge, A. (2011). Interactions between an arbuscular mycorrhizal fungus and a soil microbial community mediating litter decomposition. *FEMS Microbiology Ecology, 80*, 236–247. https://doi.org/10.1111/j.1574-6941.2011.01292.x.

Hijri, M., & Sanders, I. R. (2005). Low gene copy number shows that arbuscular mycorrhizal fungi inherit genetically different nuclei. *Nature, 433*, 160–163.

Inkpen, S. A., Douglas, G. M., Brunet, T. D. P., et al. (2017). The coupling of taxonomy and function in microbiomes. *Biology & Philosophy.* Published online 1 November 2017. https://doi.org/10.1007/s10539-017-9602-2

Jacobs, K. B., Yeager, M., Zhou, W., et al. (2012). Detectable clonal mosaicism and its relationship to aging and cancer. *Nature Genetics, 44*,651–658. http://www.nature.com/ng/journal/v44/n6/abs/ng.2270.html#supplementary-information

Jónsson, H., Sulem, P., Kehr, B., et al. (2017). *Parental influence on human germline de novo mutations in 1,548 trios from Iceland.* Nature advance online publication

Kuhn, G., Hijri, M., & Sanders, I. R. (2001). Evidence for the evolution of multiple genomes in arbuscular mycorrhizal fungi. *Nature, 414*, 745–748.

Laurie, C. C., Laurie, C. A., Rice, K., et al. (2012). Detectable clonal mosaicism from birth to old age and its relationship to cancer. *Nature Genetics, 44*, 642–650. http://www.nature.com/ng/journal/v44/n6/abs/ng.2271.html#supplementary-information

Marleau, J., Dalpe, Y., St-Arnaud, M., & Hijri, M. (2011). Spore development and nuclear inheritance in arbuscular mycorrhizal fungi. *BMC Evolutionary Biology, 11*, 51.

Matos, I., Sucena, E., Machado, M., et al. (2011). Ploidy mosaicism and allele-specific gene expression differences in the allopolyploid Squalius alburnoides. *BMC Genetics, 12*, 101.

Maynard Smith, J., & Szathmáry, E. (1995). *The major transitions in evolution.* Oxford: W. H. Freeman.

McInerney, J. O., Pisani, D., Bapteste, E., & O'Connell, M. J. (2011). The public goods hypothesis for the evolution of life on Earth. *Biology Direct, 6*, 41.

Michod, R. E., & Roze, D. (2001). Cooperation and conflict in the evolution of multicellularity. *Heredity, 86*, 1–7.

Monro, K., & Poore, A. G. B. (2004). Selection in modular organisms: Is intraclonal variation in macroalgae evolutionarily important? *The American Naturalist, 163*, 564–578. https://doi.org/10.1086/382551.

Morris, J. J., Lenski, R. E., & Zinser, E. R. (2012). The Black Queen Hypothesis: Evolution of dependencies through adaptive gene loss. *MBio, 3*, e00036–e00012.

O'Malley, M. (2014). *Philosophy of microbiology.* Cambridge: Cambridge University Press.

Okasha, S. (2006). *Evolution and the levels of selection*. Oxford: Oxford University Press.

Pineda-Krch, M., & Lehtila, K. (2004a). Challenging the genetically homogeneous individual. *Journal of Evolutionary Biology, 17*, 1192–1194.

Pineda-Krch, M., & Lehtila, K. (2004b). Costs and benefits of genetic heterogeneity within organisms. *Journal of Evolutionary Biology, 17*, 1167–1177.

Pradeu, T. (2016). The many faces of biological individuality. *Biology and Philosophy, 31*, 761–773. https://doi.org/10.1007/s10539-016-9553-z.

Queller, D. C., & Strassmann, J. E. (2009). Beyond society: The evolution of organismality. *Philosophical Transactions of the Royal Society B: Biological Sciences, 364*, 3143–3155. https://doi.org/10.1098/rstb.2009.0095.

Queller, D. C., & Strassmann, J. E. (2016). Problems of multi-species organisms: Endosymbionts to holobionts. *Biology and Philosophy, 31*, 855–873. https://doi.org/10.1007/s10539-016-9547-x.

Riley, R., Charron, P., Marton, T., & Corradi, N. (2016). Evolutionary genomics of arbuscular mycorrhizal fungi. In *Molecular mycorrhizal symbiosis* (p. 421). Hoboken: Wiley.

Rinkevich, B. (2004). Will two walk together, except they have agreed? Amos 3:3. *Journal of Evolutionary Biology, 17*, 1178–1179.

Rinkevich, B. (2005). Natural chimerism in colonial urochordates. *Journal of Experimental Marine Biology and Ecology, 322*, 93–109. https://doi.org/10.1016/j.jembe.2005.02.020.

Roger, A., Colard, A., Angelard, C., & Sanders, I. R. (2013a). Relatedness among arbuscular mycorrhizal fungi drives plant growth and intraspecific fungal coexistence. *The ISME Journal, 7*, 2137–2146.

Roger, A., Gétaz, M., Rasmann, S., & Sanders, I. R. (2013b). Identity and combinations of arbuscular mycorrhizal fungal isolates influence plant resistance and insect preference. *Ecological Entomology, 38*, 330–338. https://doi.org/10.1111/een.12022.

Samuels, M. E., & Friedman, J. M. (2015). Genetic mosaics and the germ line lineage. *Genes, 6*, 216–237.

Sanders, I. R., & Rodriguez, A. (2016). Aligning molecular studies of mycorrhizal fungal diversity with ecologically important levels of diversity in ecosystems. *The ISME Journal, 10*, 2780–2786.

Santelices, B. (1999). How many kinds of individual are there? *Trends in Ecology & Evolution, 14*, 152–155.

Santelices, B. (2004a). Mosaicism and chimerism as components of intraorganismal genetic heterogeneity. *Journal of Evolutionary Biology, 17*, 1187–1188. https://doi.org/10.1111/j.1420.9101.2004.00813.x.

Santelices, B. (2004b). A comparison of ecological responses among aclonal (unitary), clonal and coalescing macroalgae. *Journal of Experimental Marine Biology and Ecology, 300*, 31–64. https://doi.org/10.1016/j.jembe.2003.12.017.

Santelices, B., & Aedo, D. (2006). Group recruitment and early survival of Mazzaella Laminarioides. *Journal of Applied Phycology, 18*, 583–589. https://doi.org/10.1007/s10811-006-9067-1.

Santelices, B., & Alvarado, J. L. (2008). Demographic consequences of coalescence in sporeling populations of Mazaella Laminarioides (Gigartinales, Rhodophyta). *Journal of Phycology, 44*, 624–636. https://doi.org/10.1111/j.1529-8817.2008.00528.x.

Santelices, B., Correa, J. A., Aedo, D., et al. (1999). Convergent biological processes in coalescing Rhodophyta. *Journal of Phycology, 35*, 1127–1149. https://doi.org/10.1046/j.1529-8817.1999.3561127.x.

Santelices, B., Aedo, D., Hormazabal, M., & Flores, V. (2003). Field testing of inter- and intraspecific coalescence among mid-intertidal red algae. *Marine Ecology Progress Series, 250*, 91–103. https://doi.org/10.3354/meps250091.

Santelices, B., Alvarado, J. L., & Flores, V. (2010). Size increments due to interindividual fusions: How much and for how long? *Journal of Phycology, 46*, 685–692. https://doi.org/10.1111/j.1529-8817.2010.00864.x.

Santelices, B., Alvarado, J. L., Chianale, C., & Flores, V. (2011). The effects of coalescence on survival and development of Mazzaella laminarioides (Rhodophyta, Gigartinales). *Journal of Applied Phycology, 23*, 395–400. https://doi.org/10.1007/s10811-010-9566-y.

Sarkar, S. (2002). Defining "Biodiversity"; Assessing biodiversity. *The Monist, 85*, 131–155.

Schweinsberg, M., Weiss, L. C., Striewski, S., et al. (2015). More than one genotype: How common is intracolonial genetic variability in scleractinian corals? *Molecular Ecology, 24*, 2673–2685. https://doi.org/10.1111/mec.13200.

Spinner, N. B., & Conlin, L. K. (2014). Mosaicism and clinical genetics. *American Journal of Medical Genetics. Part C, Seminars in Medical Genetics, 166*, 397–405. https://doi.org/10.1002/ajmg.c.31421.

Staley, J. T., & Konopka, A. (1985). Measurement of in situ activities of nonphotosynthetic microorganisms in aquatic and terrestrial habitats. *Annual Review of Microbiology, 39*, 321–346.

Strassmann, J. E., & Queller, D. C. (2004). Genetic conflicts and intercellular heterogeneity. *Journal of Evolutionary Biology, 17*, 1189–1191.

Strassmann, J. E., & Queller, D. C. (2010). The social organism: Congresses, parties, and committees. *Evolution, 64*, 605–616. https://doi.org/10.1111/j.1558-5646.2009.00929.x.

van der Heijden, M. G., de, B. S., Luckerhoff, L., et al. (2016). A widespread plant-fungal-bacterial symbiosis promotes plant biodiversity, plant nutrition and seedling recruitment. *The ISME Journal, 10*, 389–399.

Vijg, J. (2014). Somatic mutations, genome mosaicism, cancer and aging. *Molecular Genetics Bases of Disease, 26*, 141–149. https://doi.org/10.1016/j.gde.2014.04.002.

Wagg, C., Bender, S. F., Widmer, F., & van der Heijden, M. G. A. (2014). Soil biodiversity and soil community composition determine ecosystem multifunctionality. *Proceedings of the National Academy of Sciences, 111*, 5266–5270.

Wang, B., & Qiu, Y. L. (2006). Phylogenetic distribution and evolution of mycorrhizas in land plants. *Mycorrhiza, 16*, 299–363.

Wernberg, T. (2005). Holdfast aggregation in relation to morphology, age, attachment and drag for the kelp Ecklonia radiata. *Aquatic Botany, 82*, 168–180. https://doi.org/10.1016/j.aquabot.2005.04.003.

Wilson, D. S., & Sober, E. (1994). Reintroducing group selection to the human behavioral sciences. *The Behavioral and Brain Sciences, 17*, 585–608.

Wyss, T., Masclaux, F. G., Rosikiewicz, P., et al. (2016). Population genomics reveals that within-fungus polymorphism is common and maintained in populations of the mycorrhizal fungus Rhizophagus irregularis. *The ISME Journal, 10*, 2514–2526.

Youssoufian, H., & Pyeritz, R. E. (2002). Mechanisms and consequences of somatic mosaicism in humans. *Nature Reviews Genetics, 3*, 748–758.

Zarraonaindia, I., Smith, D., & Gilbert, J. (2013). Beyond the genome: Community-level analysis of the microbial world. *Biology and Philosophy, 28*, 261–282. https://doi.org/10.1007/s10539-012-9357-8.

11

Biodiversity and Evolutionary Developmental Biology

Alessandro Minelli

Abstract A key problem in conservation biology is how to measure biological diversity. Taxic diversity (the number of species in a community or in a local biota) is not necessarily the most important aspect, if what most matters is to evaluate how the loss of the different species may impact on the future of the surviving species and communities. Alternative approaches focus on functional diversity (a measure of the distribution of the species among the different 'jobs' in the ecosystem), others on morphological disparity, still others on phylogenetic diversity. There are three major reasons to prioritize the survival of species which provide the largest contributions to the overall phylogenetic diversity. First, evolutionarily isolated lineages are frequently characterized by unique traits. Second, conserving phylogenetically diverse sets of taxa is valuable because it conserves some sort of trait diversity, itself important in so far as it helps maintain ecosystem functioning, although a strict relationships between phylogenetic diversity and functional diversity cannot be taken for granted. Third, in this way we maximize the "evolutionary potential" depending on the *evolvability* of the survivors. This suggests an approach to conservation problems focussed on evolvability, robustness and phenotypic plasticity of developmental systems in the face of natural selection: in other terms, an approach based on evolutionary developmental biology.

Keywords Evolvability · Functional diversity · Morphological disparity · Phenotypic plasticity · Phylogenetic diversity

11.1 A Concern for Biodiversity: Evolution's Products at Risk

A key problem in conservation biology is how to measure the biological diversity at risk of loss, or already lost at the global scale or in a given area or habitat.

A. Minelli (✉)
Department of Biology, University of Padova, Padova, Italy
e-mail: alessandro.minelli@unipd.it

The origin of the concept of biodiversity from within ecology (Wilson 1988) explains why biological diversity is primarily described and measured in terms of the number of species in a community, or in a local biota. However, when describing the ongoing extinction of the Anthropocene, the total number of species involved is not necessarily the most important issue. What matters in the end is what has been lost or may be lost with them, and how this loss may impact on the future of both species and communities.

What we eventually prioritize is often heavily biased by our emotional preference for a few kinds of organisms and also for selected areas or habitats. Vertebrates are given much more attention than nematodes, big cats quite more than rodents. Whole biotas of particular sites are cause of special concern, for example those of remote oceanic islands like the Hawaii or the Galápagos. In other instances, individual species become the target of dedicated conservation efforts because of the peculiar role they play in the ecosystem, for example (in the case of bees and other insects) as pollinators or (in the case of corals) as builders of reefs on which the existence of many other marine species depends. Other reasons for identifying a species as worth of special conservation effort are less obvious and perhaps, *prima facie*, just academic or antiquarian – as for example, when we decide that a given species is worth of special conservation effort only because it is the only member or the last survivor of a peculiar evolutionary lineage.

11.1.1   Beyond Species Number

The latter example deserves closer scrutiny. Low species number is not necessarily a sign of scarce success of the whole lineage, or of impending risk of extinction: in several instances, it is a consequence of ecological marginality, that is, of adaptation to infrequent habitats, or to habitats confined to very small corners of the planet's surface. Ricklefs (2005) demonstrated that this is indeed the case for a number of tribes and even families of passerine birds. In this huge group (about 6000 extant species), one to five species belonging to each of these small subgroups are morphologically quite unusual and this correlates with their adaptation to marginal habitats. For example, these birds have unusually long legs and elongated bills facilitating feeding when the birds are perching or forage on hard substrates such as bark or rock. Within each of these small groups, genetic distances among the few extant species are generally large, indicating old divergence and, by inference, low speciation rate – opposite to the trend prevailing in species-rich, successful groups inhabiting more widely distributed habitats. The fact that these groups are still around in spite of a low speciation rate suggests that in these groups also the rate of extinction is lower than the average.

Clearly, estimates of biological diversity limited to counting species number in a community or in the fauna or flora of a given area fail to capture all the information we need to obtain a satisfactory assessment of possible criticalities and, as a consequence, to formulate sensible conservation measures that might be eventually adopted.

Indeed, a number of metrics of biological diversity have been proposed (summarized in Erwin 2008) other than those that measure just *taxic diversity*, i.e. the number of species or of lower (e.g., subspecies) or higher (e.g., genera or families) classificatory units. Some alternative approaches focus instead on *functional diversity* (a measure of the distribution of the species among the different 'jobs' in the ecosystem), others on *morphological disparity*, still others on *phylogenetic diversity*. All these approaches (which should be intended as complementary rather than alternative, although their usefulness is likely to be uneven in different instances) result in estimates of biodiversity to which all species in the community or biota contribute, although not necessarily to the same extent. In the next paragraphs I will briefly comment on these metrics, before moving to a less conventional approach to conservation problems, largely based on intrinsic properties of the individual species.

11.1.2   Disparity vs. Diversity

In terms of species number, birds are more diverse than mammals (some 10,000 vs. ca. 5600 extant species worldwide), but are instead quite more uniform in terms of morphology, reproductive biology and developmental schedules. Even including less conventional kinds such as the flightless penguins and ratites (ostrich and relatives), the range of bird structural types is much narrower than the range of structural types of mammals, which include humans and whales, bats and giraffes, moles and armadillos. All birds are oviparous, whereas in mammals there are a handful of oviparous species alongside a vast majority of viviparous species. Among the latter, some, like kangaroos, are borne at a developmental stage that deserves be called a larva, whereas others develop in their mother's wombs up to a much more advanced stage and are often capable to feed for themselves in the very day in which they are born. In technical terms, the *disparity* of mammals is much higher than the disparity of birds.

The choice of characters we can consider to evaluate a group's disparity is arbitrary, but morphological traits are usually given priority, often exclusive (Foote 1997; McGhee 1999; Wills 2001; Erwin 2007), because these aspects are the most readily accessible to quantification. As noted by Gerber et al. (2008), the concept of morphological disparity (Gould 1989, 1991; Foote 1997) has proved to be an invaluable source of information, both in palaeontology (e.g., Foote 1993, 1995, 1997; Wills et al. 1994; Dommergues et al. 1996; Roy and Foote 1997) and in the study of extant organisms (e.g., Ricklefs and Miles 1994; Ricklefs 2005).

In some lineages, the level of disparity goes together with the success as measured in terms of species diversity. Examples are some huge animal and plant genera among whose representatives disparate body plans or life styles have evolved in a relatively short time. Examples include *Megaselia* (a genus of phorid flies of which 1559 species have been described, but these are probably a minor subset of those existing on Earth, and their morphological and ecological disparity are enormous); among the flowering plants, genera combining high diversity and high disparity include *Euphorbia* (2150 species, ranging from tiny herbs to quite large trees, and also including a number of succulents) and *Lobelia* (417 species, among which are small herbaceous plants alongside woody giants). However, there are also many large animal and plant genera within which the morphological differences are minor (low disparity), and vice versa (Minelli 2016). An example of low diversity combined with high disparity is the phylum Ctenophora, with 165 species described thus far, classified in 27 families, ten of which include only one species each.

11.1.3 Functional Diversity

By measuring disparity rather than simply counting species, we move a step in the direction of acknowledging the different functional roles the individual species play in respect to their biotic and abiotic environment. This aspects has been addressed in a more direct way by a number of approaches to biodiversity which try to capture the so-called *functional diversity*, the component of diversity that influences ecosystem dynamics, stability, productivity, nutrient balance, and other aspect of ecosystem functioning (Tilman 2001) through targeted descriptors and the calculation of corresponding indices (e.g., Mason et al. 2005; Bremner 2008; Villéger et al. 2008; Laliberté and Legendre 2010; Schleuter et al. 2010; Mouillot et al. 2013; Gagic et al. 2015; Gusmao-Junior and Lana 2015). Estimates of functional diversity are based, for example, on the number and kinds of trophic groups (e.g., primary producers, primary consumers, predators, parasites) and the number and relative abundance of species belonging to each group.

11.1.4 Phylogeny vs. Function

Phylogeny, and evolution at large, feature prominently in assessments of biological diversity and disparity, but it is not always obvious why. Of course, evolution is responsible both for the origin of the species whose number is the most popular measure of biodiversity, and for their structural and functional disparity, the two aspects mirroring the two main facets of evolutionary process – the splitting of lineages (cladogenesis) and the steady changes accumulating along each lineage (anagenesis), respectively. However, the frequent focus on phylogenetic diversity as an estimate of biodiversity and a criterion for ranking species to establish conservation priorities (Buckley 2016), deserves some explanation.

Several algorithms have been proposed to calculate phylogenetic diversity, based on the cladistic relations among the taxa (more often species, but also infraspecific units) (e.g., Vanewright et al. 1991; Faith 1992; Crozier 1997; Moritz 2002; Tucker et al. 2017).

Following the work of Vane-Wright et al. (1991), Faith (1992) and later authors (summarized in Mazel et al. 2017), there are three major reasons to prioritize the survival of species representing species-poor lineages only distantly related to the others in the sample and thus providing the largest contributions to the overall phylogenetic diversity.

First, in this way we conserve the greatest amount of evolutionary history (Vane-Wright et al. 1991), an ill-defined concept at the core of which, however, there is a sensible notion: evolutionarily isolated lineages, often represented by only one or very few extant species, are frequently characterized by unique traits, such as the two continuously growing leaves of *Welwitschia* or the egg-laying habit of the monotremes, strongly contrasting with the viviparity of all other mammals (Rosauer and Mooers 2013).

Second, conserving phylogenetically diverse sets of taxa is valuable because it conserves some sort of trait diversity (e.g., Mazel et al. 2017) This is the most popular among the arguments advocated in favour of using phylogenetic diversity as a basis on which to determine priorities for conservation. Trait diversity is considered important in so far as it helps maintain ecosystem functioning (e.g., Cadotte et al. 2008; Best et al. 2013; Winter et al. 2013; Gross et al. 2017).

Unfortunately, a strict relationships between phylogenetic diversity and functional diversity cannot be taken for granted. Through an elegant set of mathematical simulations, Mazel et al. (2017) were able to demonstrate that basing on estimates of phylogenetic diversity a strategy for conserving functional diversity is not necessarily a good strategy: the relationships between these two aspects of diversity depend on the shape of the tree depicting the phylogenetic relationships among the species involved and also on the model according to which their traits evolve across the generations. Therefore, generalizations are unwarranted. Still worse, Mazel et al. (2017) found that under plausible scenarios of evolution and ecology, prioritizing species conservation based on phylogenetic diversity can actually lead to levels of functional diversity lower than those obtained by conservation priorities determined by a random listing of the same species.

The third reason often advocated to explain why conservation priorities should be based on phylogenetic diversity is that in this way we maximize the "evolutionary potential" of the surviving biota (Faith 1992; Forest et al. 2007). As explained in a later section of this article, this vague term acquires a precise meaning and content when approached from the point of view of evolutionary developmental biology.

11.1.5 Antiquarian Sensibility

We value some human artefacts because of their current usefulness or at least because of the aesthetic pleasure we obtain by looking at them; but we also value other artefacts, even if devoid of any practical use and aesthetic qualities, simply because of their age. Museums are full of nondescript pieces of metal, bone or stone, witnesses of the human presence in particular sites at particular and often remote times, and of the cultural evolution of our ancestors. Similarly, a plant or animal lineage is often regarded as one of singular conservation value just because of the very long time since it split away from its closest living relatives. Monotremes (of which the platypus and the echidnas are the only living representatives) are an obvious example: the last common ancestor they share with the other living mammals lived between 162.9 and 191.1 million years ago (dos Reis et al. 2012). This circumstance, together with the strong unbalance in species richness (5 species only in the monotremes, compared to more than thousand times as many in the sister branch–the Theria, that is marsupials plus placentals) provides a good argument for regarding the platypus and the echidnas as a group of mammals we should not risk to bring to extinction. Another example is the tuatara, a reptile quite similar to a large lizard, but anatomically peculiar enough to deserve being placed in a distinct order, the Rhynchocephalia, of which it is the only survivor, confined to about 30 small islands off the North Island of New Zealand. This group separated from the Squamata (lizards, snakes and allies) about 228 million years ago or earlier (Hipsley et al. 2009).

Sometimes we realize too late the amount of history that is cancelled with the extinction of the last survivor(s) of a plant or animal group. This happened for example with the nesophontids, small mammals of which eight different species inhabited Cuba, Hispaniola, Puerto Rico and the Cayman Islands until their recent extinction, probably caused by the introduction of black rats by European sailors ca. 500 years ago. The nesophontids are classified with the insectivores and their closest relatives are the solenodontids, also confined in the Caribbean area. The two living species of the latter family, however, are poor substitutes for the loss of the nesophontids, not simply because they are themselves on the verge of extinction, but especially because the split between the two families (Nesophontidae and Solenodontidae) is very old, more than 50 million years (Brace et al. 2016) – longer, for example, than the age of the split between the New Worlds monkeys (the platyrrhines) and the Old World monkeys, including apes and humans (the catarrhines), and broadly the same age as the split between the ruminants and the lineage including hippos, whales and dolphins (O'Leary et al. 2013).

11.2   Conserving Evolutionary Processes

As noted by Buckley (2016), "by conserving genetic or phylogenetic diversity, we are facilitating the ability of lineages to adapt to future environmental changes." Since the early times of what eventually became conservation biology, far-seeing scientists have remarked that strategies for the long-term survival of wild species must take into account the continuing evolution of populations: as a consequence, policies should be based on adequate understanding of the population-genetics principles of conservation (Frankel 1974) about which quite little was known at the time. Twenty years later, progress in this direction was still insufficient, witness the plea of Smith et al. (1993, p. 164) who stressed that "If we are to conserve biodiversity the ecological and evolutionary mechanisms generating genetic diversity and the isolating mechanisms critical for speciation must also be preserved." Things have not changed much in the following years, and Moritz (2002) still lamented that "Less progress has been made on how to prioritize habitats, species, or populations in relation to persistence, that is, ensuring that the processes that sustain current and future diversity are protected."

Some authors (e.g., Gillson 2015) have remarked the paradox of conservation: we seek to preserve systems that are incessantly in flux, because of a number of processes running at different spatial and temporal scale, partly driven by extrinsic factors such as climate change and human disturbancy, partly expressing the organisms' evolutionary dynamics that would be innatural to contrast, if ever it would prove possible. Thus, if we can try to contrast the current biodiversity crisis by limiting the human impact on the environment, and even try to reduce, at least in some areas, the disruptive effect of rapid climate change, we may better help the survival of living species and lineages by devising conservation policies based on a sound understanding of *evolvability*.

11.3   Evo-Devo: Evolvability, Robustness, Plasticity

What is evolvability? Unfortunately, this is one of those technical terms for which too many definitions have been proposed (Pigliucci 2008; Brookfield 2009; Minelli 2017). Most of these, however, agree on regarding evolvability as the ability of populations to generate heritable phenotypic variation (Brigandt 2007; Kirschner and Gerhart 1998; G. P. Wagner and Altenberg 1996), but some are quite more specific, e.g. in focussing on the capacity to evolve new adaptations (Bedau and Packard 1992). Eventually, I prefer the definition proposed by Masel and Trotter (2010, p. 406), according to which evolvability is "the capacity of a population to produce heritable phenotypic variation of a kind that is not unconditionally deleterious. This definition includes both evolution from standing variation and the ability of the population to produce new variants."

According to Hendrikse et al. (2007), focussing on evolvability is the most characteristic feature of the research programme of *evolutionary developmental biology* (also called evo-devo). This young branch of the life sciences has much to offer to conservation (Campbell et al. 2017). Up to now, conservation efforts based on the preservation of genetic variation have followed the approach to intraspecific diversity characteristic of population genetics. But this is too limiting: as remarked long ago by Waddington (1957), the expression of genetic variation is structured by development. And this is exactly where evo-devo operates, in a systematic effort to unravel the complex relationships between genotype and phenotype (the so-called genotype→phenotype map; cf. Alberch 1991; Wagner and Altenberg 1996; Pigliucci 2001; West-Eberhard 2003; Draghi and Wagner 2008).

Indeed, to understand evolvability, we must acknowledge that the path leading from genotype to phenotype is complex and not necessarily predictable (Minelli

2017). On the one hand, due to environmental influences but also to stochastic impredictability, different phenotypes can be produced by developing organisms that share identical genotypes; reciprocally, identical phenotypes can be produced by developmental systems with different genotypes. Elaborating upon Waddington's insight, students of evo-devo have demonstrated that the expression of genetic variation is largely dependent on the structure and *robustness* of the developing system (Hansen 2006; Kirschner and Gerhart 1998; Wagner 2005; Wagner and Altenberg 1996).

Together with evolvability, robustness plays a central role in evolutionary developmental biology. The robustness of a phenotypic trait can be operationally defined as the absence of variation in the face of environmental or genetic perturbations (Félix and Barkoulas 2015). According to some authors, robustness constrains and contrasts evolvability, with negative effects on biodiversity: the rationale is that, if mutations and environmental changes have little effect, there is not much variation on which selection can act. Others (e.g., Kitano 2004; Wagner 2008; Masel and Trotter 2010; Melzer and Theißen 2016; Theißen and Melzer 2016) regard this view as simplistic and even contend that robustness may promote evolvability, i.e. the ability to produce heritable phenotypic variation (Pigliucci 2008). To explain how, we must first distinguish between two aspects of robustness, genetic vs. developmental.

Genetic robustness is "robustness to perturbations both in the form of new mutations and in the form of the creation of new combinations of existing alleles by recombination" (Masel and Trotter 2010, p. 407), without visible effects on the phenotype. In this way, in the absence of exposure to novel selective challenges, populations accumulate genetic diversity on the base of which they gain easier access to a greater range of novel genotypes, some of which may eventually prove to be advantageous (A. Wagner 2005, 2008, 2011).

A similar relationships between robustness and evolvability is found in the case of developmental robustness (also known as *canalization*, a term coined by Waddington (1942)) that is, the production of the same phenotype irrespective of genetic differences (and external perturbations). This also corresponds to the fact that populations harbour amounts of unexpressed genetic variation (*cryptic genetic variation*) that is not expressed in the phenotype unless revealed by environmental change or by modification in the overall genetic background (e.g., Badyaev 2005; Flatt 2005; Gibson and Dworkin 2004; Moczek 2007; Rieseberg et al. 2003; Schlichting 2008). This cryptic variation represents a standing potential for evolvability. Exposure to novel selective pressures can be dramatically accelerated by the human impact on the environment. In other terms, environmental change does not just alter the selective regime to which a population is exposed, but can also induce novel developmental responses even in the absence of genetic change. This property of the genotype→phenotype map is known as *phenotypic plasticity* (Fusco and Minelli 2010).

Phenotypic plasticity should not be regarded as an alternative to natural selection. On the one hand, the emergence of a novel phenotype by plasticity, following exposure to previously unexperienced environmental conditions creates a new target on which selection will operate; on the other, plasticity itself is subjected to selection, being favoured in fluctuating environments (Price et al. 2003; West-Eberhard 2003; Pfennig et al. 2010). As noted by Campbell et al. (2017), this is a situation likely experienced by a population introduced in a new area or living in habitats fragmented or otherwise damaged by man.

One might argue that phenotypic plasticity, although responsible for the emergence of new phenotypes, offers no warranty of their conservation, on the long term at least. But this would be a short-sighted perspective. An environmentally controlled phenotype can eventually fall under strict genetic control. The functional divide to be crossed is sometimes a very narrow one, as demonstrated by the control of wing development in the pea aphid, *Acyrtosiphum pisum*. In this little insect, some adults (males as well as females) are winged, while the others are wingless. In the male sex, the coexistence of these two alternative phenotypes is under genetic control, while in the females wing development is controlled by the exposure to different day-lengths in a critical phase of development. In technical terms, males exhibit genetic polymorphism for this trait, while females exhibit an environmentally controlled polyphenism, ie the outcome of phenotypic plasticity. This contrast, however, rests on minor mechanistic differences, because the developmental effect of day-length on wing development in the females is mediated by the gene product of the same gene whose alternative alleles are responsible for the wing polymorphism in the male (Braendle et al. 2005a, b). This circumstance suggests how easily a polyphenism can evolve into a genetic polymorphism, eventually allowing long-term conservation of phenotypes.

11.4   A Lesson from Past Mass Extinctions?

Irrespective of the different causes involved in these events, the mass extinctions of the past should be studied very carefully by researchers interested in conservation biology, but not so much to analyze the differential tribute paid by organisms belonging to different lineages, as to look for any possible explanation of the differential success of the survivors in the post-crisis recovery. Palaeobiologists have generally focussed on the ecological determinants of this process; that is, they have regarded the ecological space left empty by extinctions as the main determinant of the renewed occupation of morphospace. To some extent, the morphological disparity often expanded into dimensions other than those that were occupied prior to the mass extinction. However, no really new body plan emerged. This was, in a sense, a large-scale test demonstrating the developmental robustness of the main traits of body architecture of the survivors, the innovations being limited to secondary, evolutionarily plastic aspects (Erwin 2008).

Confronted with this (admittedly, only incompletely documented) evidence, it seems legitimate to rethink the evolutionary criteria in the light of which biodiversity and its ongoing loss are currently evaluated. It is hard to imagine a positive correlation between the phylogenetic relationships among the survivors and the possible outcome of their long-term evolution in a post-crisis recovery. Million years ago, by preferring to save a marsupial and a placental mammal rather than two placentals, because of the larger phylogenetic distance between the first two, we would not have been able to predict that at least two different subterranean lineages would have eventually evolved in any case: today there are indeed marsupial moles (*Notoryctes*) among the marsupials and moles (*Talpa*) and mole-rats (*Spalax*) among the placentals. More than because of the history of their lineage, survivors may be differentially important for the future of biodiversity as a function of their intrinsic qualities, particularly those expressed by the parameters on which evo-devo focuses – as said, robustness and evolvability.

Acknowledgements I am grateful to Elena Casetta and Davide Vecchi for their invitation to contribute to this volume and for their precious comments on a first draft text.

References

Alberch, P. (1991). From genes to phenotype: Dynamical systems and evolvability. *Genetica, 84*, 5–11.

Badyaev, A. V. (2005). Stress-induced variation in evolution: From behavioural plasticity to genetic assimilation. *Proceedings of the Royal Society of London. Series B, Biological Sciences, 272*, 877–886. https://doi.org/10.1098/rspb.2004.3045.

Bedau, M., & Packard, N. (1992). Measurement of evolutionary activity, teleology, and life. In C. Langton, C. Taylor, J. D. Farmer, & S. Rasmussen (Eds.), *Artificial life II: Proceedings of the workshop on artificial life* (pp. 431–462). Redwood City: Addison-Wesley.

Best, R. J., Caulk, N. C., & Stachowicz, J. J. (2013). Trait vs. phylogenetic diversity as predictors of competition and community composition in herbivorous marine amphipods. *Ecology Letters, 16*, 72–80. https://doi.org/10.1111/ele.12016.

Brace, S., Thomas, J. A., Dalén, L., Burger, J., MacPhee, R. D. E., Barnes, I., & Turvey, S. T. (2016). Evolutionary history of the Nesophontidae, the last unplaced recent mammal family. *Molecular Biology and Evolution, 33*, 3095–3103. https://doi.org/10.1093/molbev/msw186.

Braendle, C., Caillaud, M. C., & Stern, D. L. (2005a). Genetic mapping of aphicarus: A sex-linked locus controlling a wing polymorphism in the pea aphid (*Acyrthosiphon pisum*). *Heredity, 94*, 435–442. https://doi.org/10.1038/sj.hdy.6800633.

Braendle, C., Friebe, I., Caillaud, M. C., & Stern, D. L. (2005b). Genetic variation for an aphid wing polyphenism is genetically linked to a naturally occurring wing polymorphism. *Proceedings of the Royal Society B: Biological Sciences, 272*, 657–664. https://doi.org/10.1098/rspb.2004.2995.

Bremner, J. (2008). Species' traits and ecological functioning in marine conservation and management. *Journal of Experimental Marine Biology and Ecology, 366*, 37–47. https://doi.org/10.1016/j.jembe.2008.07.007.

Brigandt, I. (2007). Typology now: Homology and developmental constraints explain evolvability. *Biology and Philosophy, 22*, 709–725. https://doi.org/10.1007/s10539-007-9089-3.

Brookfield, J. F. Y. (2009). Evolution and evolvability: Celebrating Darwin 200. *Biology Letters, 5*, 44–46. https://doi.org/10.1098/rsbl.2008.0639.

Buckley, T. R. (2016). Applications of phylogenetics to solve practical problems in insect conservation. *Current Opinion in Insect Science, 18*, 35–39. https://doi.org/10.1016/j.cois.2016.09.005.

Cadotte, M. W., Cardinale, B. J., & Oakley, T. H. (2008). Evolutionary history and the effect of biodiversity on plant productivity. *Proceedings of the National Academy of Sciences of the United States of America, 105*, 17012–17017. https://doi.org/10.1073/pnas.0805962105.

Campbell, C. S., Adams, C. E., Bean, C. W., & Parsons, K. J. (2017). Conservation evo-devo: Preserving biodiversity by understanding its origins. *Trends in Ecology and Evolution, 32*, 746–759. https://doi.org/10.1016/j.tree.2017.07.002.

Crozier, R. H. (1997). Preserving the information content of species: Genetic diversity, phylogeny, and conservation worth. *Annual Review of Ecology and Systematics, 28*, 243–268. https://doi.org/10.1146/annurev.ecolsys.28.1.243.

Dommergues, J.-L., Laurin, B., & Meister, C. (1996). Evolution of ammonoid morphospace during the early Jurassic radiation. *Paleobiology, 22*, 219–240. https://doi.org/10.1017/S0094837300016183.

dos Reis, M., Inoue, J., Hasegawa, M., Asher, R. J., Donoghue, P. C. J., & Yang, Z. (2012). Phylogenomic datasets provide both precision and accuracy in estimating the timescale of placental mammal phylogeny. *Proceedings of the Royal Society of London. Series B, Biological Sciences, 279*, 3491–3500. https://doi.org/10.1098/rspb.2012.0683.

Draghi, J., & Wagner, G. P. (2008). Evolution of evolvability in a developmental model. *Evolution, 62*, 301–315. https://doi.org/10.1111/j.1558-5646.2007.00303.x.

Erwin, D. H. (2007). Disparity: Morphologic pattern and developmental context. *Palaeontology, 50*, 57–73. https://doi.org/10.1111/j.1475-4983.2006.00614.x.

Erwin, D. H. (2008). Extinction as the loss of evolutionary history. *Proceedings of the National Academy of Sciences of the United States of America, 105*(Suppl 1), 11520–11527. https://doi.org/10.1073/pnas.0801913105.

Faith, D. P. (1992). Conservation evaluation and phylogenetic diversity. *Biological Conservation, 61*, 1–10. https://doi.org/10.1016/0006-3207(92)91201-3.

Félix, M.-A., & Barkoulas, M. (2015). Pervasive robustness in biological systems. *Nature Reviews Genetics, 16*, 483–496. https://doi.org/10.1038/nrg3949.

Flatt, T. (2005). The evolutionary genetics of canalization. *Quarterly Review of Biology, 80*, 287–316. https://doi.org/10.1086/432265.

Foote, M. (1993). Discordance and concordance between morphological and taxonomic diversity. *Paleobiology, 19*, 185–204. https://doi.org/10.1017/S0094837300015864.

Foote, M. (1995). Morphological diversification of Paleozoic crinoids. *Paleobiology, 21*, 273–299. https://www.jstor.org/stable/2401167.

Foote, M. (1997). The evolution of morphological diversity. *Annual Review of Ecology and Systematics, 28*, 129–152. https://doi.org/10.1146/annurev.ecolsys.28.1.129.

Forest, F., Grenyer, R., Rouget, M., Davies, T. J., Cowling, R. M., Faith, D. P., Balmford, A., Manning, J. C., Procheş, Ş., van der Bank, M., Reeves, G., Hedderson, T. A. J., & Savolainen, V. (2007). Preserving the evolutionary potential of floras in biodiversity hotspots. *Nature, 445*, 757–760. https://doi.org/10.1038/nature05587.

Frankel, O. H. (1974). Genetic conservation: Our evolutionary responsibility. *Genetics, 78*, 53–65.

Fusco, G., & Minelli, A. (2010). Phenotypic plasticity in development and evolution: Facts and concepts. *Philosophical Transactions of the Royal Society of London. Series B, Biological Sciences, 365*, 547–556. https://doi.org/10.1098/rstb.2009.0267.

Gagic, V., Bartomeus, I., Jonsson, T., Taylor, A., Winqvist, C., Fischer, C., Slade, E. M., Steffan-Dewenter, I., Emmerson, M., Potts, S. G., Tscharntke, T., Weisser, W., & Bommarco, R. (2015). Functional identity and diversity of animals predict ecosystem functioning better than species-based indices. *Proceedings of the Royal Society of London. Series B, Biological Sciences, 282*, 20142620. https://doi.org/10.1098/rspb.2014.2620.

Gerber, S., Eble, G. J., & Neige, P. (2008). Allometric space and allometric disparity: A developmental perspective in the macroevolutionary analysis of morphological disparity. *Evolution, 62*, 1450–1457. https://doi.org/10.1111/j.1558-5646.2008.00370.x.

Gibson, G., & Dworkin, I. (2004). Uncovering cryptic genetic variation. *Nature Reviews Genetics, 5*, 681–691. https://doi.org/10.1038/nrg1426.

Gillson, L. (2015). *Biodiversity conservation and environmental change: Using palaeoecology to manage dynamic landscapes in the Anthropocene*. Oxford: Oxford University Press.

Gould, S. J. (1989). *Wonderful life*. New York: Norton.

Gould, S. J. (1991). The disparity of the burgess shale arthropod fauna and the limits of cladistic analysis: Why we must strive to quantify morphospace. *Paleobiology, 17*, 411–423. https://www.jstor.org/stable/2400754.

Gross, N., Le Bagousse-Pinguet, Y., Liancourt, P., Berdugo, M., Gotell, N. J., & Maestre, F. T. (2017). Functional trait diversity maximizes ecosystem multifunctionality. *Nature Ecology & Evolution, 1*, 132. https://doi.org/10.1038/s41559-017-0132.

Gusmao-Junior, J. B. L., & Lana, P. C. (2015). Spatial variability of the infauna adjacent to inter-tidal rocky shores in a subtropical estuary. *Hydrobiologia, 743*, 53–64. https://doi.org/10.1007/s10750-014-2004-4.

Hansen, T. F. (2006). The evolution of genetic architecture. *Annual Review of Ecology and Systematics, 37*, 123–157. https://doi.org/10.1146/annurev.ecolsys.37.091305.110224.

Hendrikse, J. L., Parsons, T. E., & Hallgrímsson, B. (2007). Evolvability as the proper focus of evolutionary developmental biology. *Evolution & Development, 9*, 393–401. https://doi.org/10.1111/j.1525-142X.2007.00176.x.

Hipsley, C. A., Himmelmann, L., Metzler, D., & Mueller, J. (2009). Integration of Bayesian molecular clock methods and fossil-based soft bounds reveals Early Cenozoic origin of African lacertid lizards. *BMC Evolutionary Biology, 9*, 1–13. https://doi.org/10.1186/1471-2148-9-151.

Kirschner, M., & Gerhart, J. (1998). Evolvability. *Proceedings of the National Academy of Sciences of the United States of America, 95*, 8420–8427. https://doi.org/10.1073/pnas.95.15.8420.

Kitano, H. (2004). Biological robustness. *Nature Reviews Genetics, 5*, 826–837. https://doi.org/10.1038/nrg1471.

Laliberté, E., & Legendre, P. (2010). A distance-based framework for measuring functional diversity from multiple traits. *Ecology, 91*, 299–305. https://doi.org/10.1890/08-2244.1.

Masel, J., & Trotter, M. V. (2010). Robustness and evolvability. *Trends in Genetics, 26*, 406–414. https://doi.org/10.1016/j.tig.2010.06.002.

Mason, N. W. H., Mouillot, D., Lee, W. G., & Wilson, J. B. (2005). Functional richness, functional evenness and functional divergence: The primary components of functional diversity. *Oikos, 111*, 112–118. https://doi.org/10.1111/j.0030-1299.2005.13886.x.

Mazel, F., Mooers, A. O., Dalla Riva, V., & Pennell, M. V. (2017). Conserving phylogenetic diversity can be a poor strategy for conserving functional diversity. *Systematic Biology, 66*, 1019–1027. https://doi.org/10.1093/sysbio/syx054.

McGhee, G. R. (1999). *Theoretical morphology: The concept and its applications*. New York: Columbia University Press.

Melzer, R., & Theißen, G. (2016). The significance of developmental robustness for species diversity. *Annals of Botany, 117*, 725–732. https://doi.org/10.1093/aob/mcw018.

Minelli, A. (2016). Species diversity vs. morphological disparity in the light of evolutionary developmental biology. *Annals of Botany, 117*, 795–809. https://doi.org/10.1093/aob/mcv134.

Minelli, A. (2017). Evolvability and its evolvability. In P. Huneman & D. Walsh (Eds.), *Challenges to evolutionary theory: Development, inheritance and adaptation* (pp. 211–238). New York: Oxford University Press.

Moczek, A. P. (2007). Developmental capacitance, genetic accommodation, and adaptive evolution. *Evolution & Development, 9*, 299–305. https://doi.org/10.1111/j.1525-142X.2007.00162.x.

Moritz, C. (2002). Strategies to protect biological diversity and the evolutionary processes that sustain it. *Systematic Biology, 51*, 238–254. https://doi.org/10.1080/10635150252899752.

Mouillot, D., Graham, N. A. J., Villéger, S., Mason, N. W. H., & Bellwood, D. R. (2013). A functional approach reveals community responses to disturbances. *Trends in Ecology and Evolution, 28*, 167–177. https://doi.org/10.1016/j.tree.2012.10.004.

O'Leary, M. A., Bloch, J. I., Flynn, J. J., Gaudin, T. J., Giallombardo, A., Giannini, N. P., Goldberg, S. L., Kraatz, B. P., Luo, Z. X., Meng, J., Ni, X., Novacek, M. J., Perini, F. A., Randall, Z. S., Rougier, G. W., Sargis, E. J., Silcox, M. T., Simmons, N. B., Spaulding, M.,

Velazco, P. M., Weksler, M., Wible, J. R., & Cirranello, A. L. (2013). The placental mammal ancestor and the post-K-Pg radiation of placentals. *Science, 339*, 662–667. https://doi.org/10.1126/science.1229237.

Pfennig, D. W., Wund, M. A., Snell-Rood, E. C., Cruickshank, T., Schlichting, C. D., & Moczek, A. P. (2010). Phenotypic plasticity's impacts on diversification and speciation. *Trends in Ecology and Evolution, 25*, 459–467. https://doi.org/10.1016/j.tree.2010.05.006.

Pigliucci, M. (2001). *Phenotypic plasticity: Beyond nature and nurture*. Baltimore: John Hopkins University Press.

Pigliucci, M. (2008). Is evolvability evolvable? *Nature Reviews Genetics, 9*, 75–82. https://doi.org/10.1038/nrg2278.

Price, T. D., Qvarnström, A., & Irwin, D. E. (2003). The role of phenotypic plasticity in driving genetic evolution. *Proceedings of the Royal Society of London. Series B, Biological Sciences, 270*, 1433–1440. https://doi.org/10.1098/rspb.2003.2372.

Ricklefs, R. E. (2005). Small clades at the periphery of passerine morphological space. *American Naturalist, 165*, 651–659. https://doi.org/10.1086/429676.

Ricklefs, R. E., & Miles, D. B. (1994). Ecological and evolutionary inferences from morphology: an ecological perspective. In P. C. Wainwright & S. M. Reilly (Eds.), *Ecological morphology* (pp. 13–41). Chicago/London: University of Chicago Press.

Rieseberg, L. H., Raymond, O., Rosenthal, D. M., Lai, Z., Livingstone, K., Nakazato, T., Durphy, J. L., Schwarzbach, A. E., Donovan, L. A., & Lexer, C. (2003). Major ecological transitions in wild sunflowers facilitated by hybridization. *Science, 301*, 1211–1216. https://doi.org/10.1126/science.1086949.

Rosauer, D. F., & Mooers, A. (2013). Nurturing the use of evolutionary diversity in nature conservation. *Trends in Ecology and Evolution, 28*, 322–323. https://doi.org/10.1016/j.tree.2013.01.014.

Roy, K., & Foote, M. (1997). Morphological approaches to measuring biodiversity. *Trends in Ecology and Evolution, 12*, 277–281.

Schleuter, D., Daufresne, M., Massol, F., & Argillier, C. (2010). A user's guide to functional diversity indices. *Ecological Monographs, 80*, 469–484. https://doi.org/10.1890/08-2225.1.

Schlichting, C. D. (2008). Hidden reaction norms, cryptic genetic variation, and evolvability. *Annals of the New York Academy of Sciences, 1133*, 187–203. https://doi.org/10.1196/annals.1438.010.

Smith, T. B., Bruford, M. W., & Wayne, R. K. (1993). The preservation of process: The missing element of conservation programs. *Biodiversity Letters, 1*, 164–167. https://www.jstor.org/stable/2999740

Theißen, G., & Melzer, R. (2016). Robust views on plasticity and biodiversity. *Annals of Botany, 117*, 693–697. https://doi.org/10.1093/aob/mcw066.

Tilman, D. (2001). Functional diversity. In S. A. Levin (Ed.), *Encyclopaedia of biodiversity* (pp. 109–120). San Diego: Academic.

Tucker, C. M., Cadotte, M. W., Carvalho, S. B., Davies, T. J., Ferrier, S., Fritz, S. A., Grenyer, R., Helmus, M. R., Jin, L. S., Mooers, A. O., Pavoine, S., Purschke, O., Redding, D. W., Rosauer, D. F., Winter, M., & Mazel, F. (2017). A guide to phylogenetic metrics for conservation, community ecology and macroecology. *Biological Reviews, 92*, 698–715. https://doi.org/10.1111/brv.12252.

Vane-Wright, R. I., Humphries, C. J., & Williams, P. H. (1991). What to protect – Systematics and the agony of choice. *Biological Conservation, 55*, 235–254. https://doi.org/10.1016/0006-3207(91)90030-D.

Villéger, S., Mason, N., & Mouillot, D. (2008). New multidimensional functional diversity indices for a multifaceted framework in functional ecology. *Ecology, 89*, 2290–2301.

Waddington, C. H. (1942). Canalization of development and the inheritance of acquired characters. *Nature, 150*, 563–565.

Waddington, C. H. (1957). *The strategy of the genes: A discussion of some aspects of theoretical biology*. London: Allen & Unwin.

Wagner, A. (2005). *Robustness and evolvability in living systems*. Princeton: Princeton University Press.

Wagner, A. (2008). Robustness and evolvability: A paradox resolved. *Proceedings of the Royal Society of London. Series B, Biological Sciences, 275*, 91–100. https://doi.org/10.1098/rspb.2007.1137.

Wagner, A. (2011). *The origins of evolutionary innovations: A theory of transformative change in living systems*. Oxford: Oxford University Press.

Wagner, G. P., & Altenberg, L. (1996). Complex adaptations and evolution of evolvability. *Evolution, 50*, 967–976. https://doi.org/10.1111/j.1558-5646.1996.tb02339.x.

West-Eberhard, M. J. (2003). *Developmental plasticity and evolution*. New York: Oxford University Press.

Wills, M. A. (2001). Morphological disparity: A primer. In J. M. Adrain, G. D. Edgecombe, & B. S. Lieberman (Eds.), *Fossils, phylogeny, and form* (pp. 55–144). New York: Kluwer/Plenum.

Wills, M. A., Briggs, D. E. G., & Fortey, R. A. (1994). Disparity as an evolutionary index: A comparison of Cambrian and recent arthropods. *Paleobiology, 20*, 93–130. https://doi.org/10.1017/S009483730001263X.

Wilson, E. O. (Ed.). (1988). *Biodiversity*. Washington, DC: National Academy Press.

Winter, M., Devictor, V., & Schweiger, O. (2013). Phylogenetic diversity and nature conservation: Where are we? *Trends in Ecology and Evolution, 28*, 199–204. https://doi.org/10.1016/j.tree.2012.10.015.

12

Evolutionary Significant Units and Plastic Population Species

Davide Vecchi and Rob Mills

Abstract The history of biology has been characterised by a strong emphasis on the identification of entities (e.g., macromolecules, cells, organisms, species) as fundamental units of our classificatory system. The biological hierarchy can be divided into a series of compositional levels complementing the physical and chemical hierarchy. Given this state of affairs, it is not surprising that biodiversity studies have focused on a "holy trinity" of entities, namely genes, species and ecosystems. In this chapter, we endorse the view that a process-based approach should integrate an entity-based one. The rationale of our endorsement is that a focus on entities does not address whether biological processes have the capacity to create novel, salient units of biodiversity. This alternative focus might therefore have implications for conservation biology. In order to show the relevance of process-based approaches to biodiversity, in this chapter we shall focus on a particular process: phenotypic plasticity. Specifically, we shall describe a model of plasticity that might have implications for how we conceptualise biodiversity units. The hypothesis we want to test is whether plastic subpopulations that have enhanced evolutionary potential vis a vis non-plastic subpopulations make them amenable to evolutionarily significant units (i.e., ESU) status. An understanding of the mechanisms that influence organismic evolution, particularly when under environmental stress, may shed light on the natural "conservability" capacities of populations. We use an abstract computational model that couples plasticity and genetic mutation to investigate how plasticity processes (through the Baldwin effect) can improve the adaptability of a population

D. Vecchi (✉)
Centro de Filosofia das Ciências, Departamento de História e Filosofia das Ciências,
Faculdade de Ciências, Universidade de Lisboa, Lisboa, Portugal
e-mail: dvecchi@fc.ul.pt

R. Mills
BioISI – Biosystems & Integrative Sciences Institute, Faculdade de Ciências,
Universidade de Lisboa, Lisbon, Portugal
e-mail: rob.mills@fc.ul.pt

when faced with novel environmental challenges. We find that there exist circumstances under which plasticity improves adaptability, where multi-locus fitness valleys exist that are uncrossable by non-plastic populations; and the differences in the capacity to adapt between plastic and non-plastic populations become drastic when the environment varies at a great enough rate. If plasticity such as learning provides not only within-lifetime environmental buffering, but also enhances a population's capacity to adapt to environmental changes, this would, on the one hand, vindicate a process-based approach to biodiversity and, on the other, it would suggest a need to take into account the processes generating plasticity when considering conservation efforts.

Keywords Biological hierarchy · Process-approach to conservation ·
Evolutionarily significant unit of conservation (ESU) · Plasticity · Baldwin effect ·
Evolutionary potential

12.1   Entity-Based and Process-Based Approaches Are Complementary

Biodiversity conservation poses a set of complex conceptual and practical challenges. The metaphor of healing and the analogy of nature as a patient serve the purpose of discriminating such challenges in three groups.[1] The state of nature as a patient must be diagnosed and the damage to biodiversity estimated in order to cure it via appropriate conservation actions. But in order to do so, we need to be able to take care of the patient in the right way. That is, we need to be able to characterise nature as a patient appropriately by identifying the relevant targets of treatment and conservation action. This contribution deals primarily with this conceptual challenge by arguing that proper treatment can be achieved not by exclusively targeting entities (i.e., units of biodiversity as conservation targets such as species or sites), but by concomitantly focusing on the processes generating such entities. In order to build our case, let us first explain why science requires endorsing a *complementary entity-and-process-based approach.*

Most people would identify biological entities such as organisms and species as paradigmatic. This intuitive knowledge (or folk biology) is often eventually refined, encompassing entities that can only be observed through microscopes such as cells and macromolecules. Biological knowledge is basically about what these entities do and how they develop or evolve. The focus on entities can be justified for at least two important reasons. First of all, nature can be thought in hierarchical terms as a series of part-whole relationships where the entities-wholes at a higher level of organisation are composed of entities-parts at a lower level. Secondly, the epistemological advantage of this compositional view is that an entity-based ontology can be

---

[1] The metaphor and analogy are developed in Casetta, Marques da Silva & Vecchi, Chap. 1, in this volume.

upheld despite ignorance of the processes leading to the generation (and, conversely, the decomposition) of wholes.[2]

The basic idea beneath entity-based classification is that nature is stratified into hierarchical levels of composition, i.e., "... *hierarchical divisions of stuff (paradigmatically, but not necessarily material stuff) organized by part-whole relations, in which wholes at one level function as parts at the next (and at all higher) levels*" (Wimsatt 2007, p. 201). Consider the physical hierarchy for example; the idea is that fundamental particles (e.g., fermions – quarks and leptons – and bosons) compose hadrons (neutrons and protons), which in turn compose atoms, which in turn compose molecules, which in turn compose all other solid, liquid and gaseous molecular aggregates that we can directly observe at the mesoscopic scale. Fundamental particles are the component parts of the hadron wholes; hadrons are the component parts of the atom wholes etc. Fundamental particles, hadrons, atoms etc. are entities belonging to different compositional levels. Biological hierarchies can be analogously divided into a series of compositional levels that complement the physical and chemical hierarchies: macromolecular wholes are composed of fundamental particles, hadrons, atoms, molecular parts; cellular wholes are composed of macromolecular parts; organismal wholes are composed of cellular parts etc. Thus, compositional hierarchies seem to identify "natural" and not purely human-dependent ontological components, even though the details of any ontology remain revisable in the light of scientific advances. The upshot of all this is that, if nature is indeed stratified into hierarchical compositional levels, then, on the one hand, all entities in the universe are ultimately composed of fundamental particles and, on the other, physical, chemical and biological hierarchies must be compositionally related. The corollary of this view is that compositionality implies some kind of physical reductionism because, on the one hand, everything is composed of basic physical stuff (e.g., quarks, leptons and bosons) and, on the other, the basic physical level of the hierarchy is primitive. This kind of compositional physicalism is unproblematic in many respects, even though this does not mean that chemistry and biology should straightforwardly be reduced to physics. One reason is that the explanation of the behaviour of chemical and biological systems might require reference to properties that are not ascribable to their physical components. Additionally, this kind of compositional physicalism rests on ontological fundamentalism, that is, the not so innocent assumption that we can make sense of the idea that a fundamental physical level exists at all (Schaffer 2003). This view implies the controversial hypothesis that quarks, leptons and bosons are not composed of anything at all, that they are "atoms" of composition in Democritus's sense. However, even though compositional physicalism is somehow obvious (because all entities are merely composed of physical stuff), it is at the same time epistemologically vacuous. Let us explain why by making reference exclusively to biology.

---

[2] Of course, there exist also processes leading to the decomposition of wholes. However, the striking feature of the history of life is that it is a history of "complexification", of generation of wholes. Thanks to Sandro Minelli for suggesting this clarification.

There are many potential reasons to argue that the compositional physicalism so far characterised provides an unsatisfactory account of biological entities. One argument is that biological entities are composed of non-physical components, a position that might be called vitalism. A somehow different argument states that biological entities possess properties that are lacked by their physical parts, a position that might be called emergentism. Here we shall focus on another kind of argument against compositional physicalism. Suppose we were to produce an inventory of all the relevant parts and their properties: could we infer the properties of wholes? No, because we would also need an understanding of the "rules of composition" governing the behaviour of parts in the production of wholes. An entity-based ontology cannot provide a satisfactory account of the properties of wholes without providing information about the nature of the processes governing the interactions between their parts. The basic point is that an entity-based ontology would provide limited knowledge of nature unless it is complemented with a process-based ontology. This is because of two reasons.

The first is that we cannot understand biological entities and their behaviour without knowledge about their structure and functional properties. Take a protein, a whole composed of a variety of amino acids with a number of physical (i.e., biophysical and biochemical) properties; the conformational properties of proteins are dependent on the properties of their component amino-acids but are not properties of these components; for instance, the specificity of proteins (i.e., their capacity to bind to a particular ligand) is given by their structure, where this structure is generated by the functional interaction of the polypeptide chains (themselves composed of amino acids) and the environment. The point is that knowledge of the physical properties of the macromolecular components of a protein is not enough in order to account for the structural and functional properties of proteins and thus to explain and predict their behaviour: composition does not account for structure and function. Thus, a static entity-based ontology consisting of an inventory of the compositional properties of parts does not exhaust the relevant biological properties of wholes.

The second reason is that the criteria for the identification of relevant biological entities must make reference to processes. For instance, consider organismal development: even if the embryo is very different from the adult capable of reproducing, they are the same biological individual. The individuality of an organism is not a property of its component parts (i.e., cells) but of the whole. Perhaps it would be better to say that it is not even a property but a process. Hennig (1966, p. 65) introduced the concept of semaphoront in order to make sense of this constrained organismal changeability. The semaphoront corresponds to the individual (e.g., a biological organism) in an infinitely small time span of its life history during which it remains unchanged. The same concept can probably be applied to any entity, physical, chemical or biological alike. But the semaphoront is a fiction, a conceptual device that should not be reified in order to vindicate a pure entity-based ontology. In this deep sense, it might be argued, biological entities should ultimately be thought in terms of dynamical processes: every supposedly static and unchangeable biological entity should be merely thought of as a portion of its life history (Dupré 2012).

At the same time, even a pure process-based ontology relinquishing any reference to entities would be epistemically useless: science also strives to classify types of entities and to uncover practically useful criteria for entity identification. This is clearly the case in biology: we want to be able to say that the embryo is the same individual as the adult, that this cell is and always will be eukaryotic during its life history, that this organism belongs to a particular species etc. Thus, in a very basic sense, an entity-based and a process-based ontology cannot but be complementary.

12.2 Entity-Based Approaches to Biodiversity Are Deficient

The intuitive allure and the epistemological advantages of entity-based compositional hierarchies are reflected in the literature on biodiversity. In fact, compositional hierarchies are prominent in biodiversity studies (Angermeier and Karr 1994, p. 691). For instance, taxonomic hierarchies stratify biological nature in terms of the part-whole relationship between species, genera, families, orders, classes, phyla, kingdoms, domains and superdomain. Ecological hierarchies stratify biological nature in terms of the part-whole relationship between populations, communities, ecosystems, landscapes, biomes and biosphere. Genetic hierarchies stratify biological nature in terms of the part-whole relationship between alleles, genes, chromosomes, genomes and pangenomes. All these hierarchies capture an aspect of biological nature (i.e., taxonomic inclusiveness, ecological nestedness, genetic organisation). To each level corresponds an entity type, i.e., a unit of biodiversity. A general problem with compositional hierarches of the above kind is that the biodiversity units are not immaculately characterisable, in the sense that sometimes it is difficult to recognise a biological entity as an entity of a certain type, as a certain biodiversity unit. For instance, whether a population of organisms constitutes a certain species might be open to debate and might depend on which species concept we take into account[3]; biofilms or host-symbiont consortia might be either considered organisms or communities depending on which characterisation of organism we take into account etc.[4] Biology does not provide clear-cut and universally-accepted criteria for the identification of a certain biological entity as an entity of a certain type because biological entities develop and evolve. But one of the relevant issues in conservation biology is whether entity-based compositional hierarchies of the above kind provide a satisfactory framework to characterise the units of conservation.[5] In this section we shall suggest that they do not. In order to do so, we shall consider three issues. The first concerns the justification for the choice of

---

[3] See Reydon, Chap. 8, in this volume, on the debate concerning the nature of species.

[4] See Marques da Silva and Casetta, Chap. 9, in this volume, on this issue.

[5] This is just one of the many conceptual and practical challenges posed by conserving biodiversity. See Casetta, Marques da Silva & Vecchi, Chap. 1, in this volume.

hierarchy. The second pertains to the justification for the choice of biodiversity units. The third concerns the rationale for the exclusive focus on entities.

12.2.1   The Limits of Conservation Fundamentalism

The taxonomic, ecological and genetic hierarchies are somehow conflicting, even though they might be related at some level. One proposal is that they are cleanly related at the species-population-genome level because, as Angermeier and Karr (1994, p. 691) argued, *"... any population has a taxonomic identity (species), which is characterized by a distinct genome."* This essentialist proposal is flawed at least in the sense that it is assumed that a species-specific genome exists, while what exists is a gene-pool (i.e., the totality of the genes of a given species existing at a given time, see Mayr 1970, p. 417). The species genome is thus a statistical artefact reconstructed with reference to this gene pool and ideally comprising all the genomic constituents of all genomes of present (but not past and future) organisms belonging to a species. If the species-population-genome relationship were clean, it would follow that by saving all present members of one species we would conserve all species-specific genomic variation, which is clearly not the case at least in the sense that some genomic variants have been surely lost in the course of evolution and others will be acquired. Of course, other ways of carving nature might exist and other compositional hierarchies, possibly linking the three hierarchies used so far, might be devised. Sarkar (2002) has for instance proposed that two compositional hierarchies should be used, one spatial (i.e., biological molecules, macromolecules, organelles, cells, organisms, populations, meta-populations, communities, ecosystems, biosphere) and one taxonomic (alleles, genes, genotypes, subspecies, species, genera, etc. until kingdoms, domain and superdomain). The advantage of this proposal is its parsimony, particularly the merging of the genetic and taxonomic hierarchies (which implies that genomic units – i.e., functional or structural genomic components – should be considered taxonomic ones).

Given a multiplicity of hierarchies, is there a possible justification for choosing one particular compositional one? For reasons that will be uncovered in this section, we strongly doubt it. However, let us suppose for the sake of argument that no hierarchy can be privileged. The following question is whether some units should be chosen as fundamental units of conservation. Clearly, it is practically impossible to conserve all diversity at all levels of a hierarchy, as it is practically impossible to focus conservation effort on the all-comprehensive top-level unit (i.e., biota and biosphere). How should we choose relevant units then? As a matter of fact, conservation practice seems to bypass this foundational question. As Sarkar (2005, p. 182) relates, the convention in conservation practice is to choose the "holy trinity" of genes, species and ecosystems as units of conservation. Note that these three types of entity belong to different hierarchies, however compositional hierarchies are characterised. Thus, conservation biologists seem to think that, for instance, conserving genes is not sufficient to conserve ecosystems and vice versa. Are they correct? Let us analyse the holy trinity in detail by starting with genes.

As we have seen in the first section, compositional hierarchies betray a reductionist bias. Unlike in physics, in the life sciences this bias is not articulated as a problem concerning ontological fundamentalism (the idea that a compositional level is primitive): obviously no biological compositional level is primitive and ontologically fundamental given that all biological entities are made of physical stuff. But an analogous problem presents itself nonetheless: is there any reason to think that a particular biological compositional level is causally privileged? Usually this question is framed in terms of reduction: suppose that biological compositional level x is adopted as privileged, would it be possible to reduce all biological phenomena to interactions between entities at that level? Generally, the answer to this question has been negative, with few interesting exceptions, for instance in developmental biology (Rosenberg 1997; Wolpert 1994). Nonetheless, a tendency to consider the molecular level as the biologically privileged level is clearly present in many branches of biology. The reason is that it is thought that the behaviour of biological wholes should, in order to be properly understood, be unpacked in terms of molecular interactions. When we move to conservation practice, the related reductionist idea seems to be that genes are the fundamental unit of conservation because, by conserving all genomic variation, we concomitantly preserve much of the phenotypic variation that characterises the populations constituting the species and higher taxonomic levels. Sarkar (2002, p. 152) notes that this position can be justified only if some form of "global genetic reductionism" (i.e., the thesis according to which "all biological features are, in some significant way, reducible to the genes") is vindicated. In a very clear sense, global genetic reductionism is wrong, fundamentally because phenogenesis at all levels (from transcription, translation and protein folding up to cellular differentiation and morphogenesis) is causally influenced by a variety of environmental inputs. Thus, saving all genes would not save all possible phenotypic outcomes unless we also conserved all possible developmental environments, which verges on the impossible.[6] Interestingly, note that developmental environments (e.g., the folding environments of proteins considered in Sect. 12.1) are fundamentally ecosystems. Also note that genes are units of the taxonomic hierarchy while developmental ecosystems are units of the spatial or ecological one. This explains why a compositional hierarchy cannot be privileged over the others and why, as a matter of fact, focusing conservation efforts on units of two different hierarchies might turn out to be a necessary rather than an incoherent conservation strategy (for a similar argument, see Sarkar 2002, p. 152).

---

[6] There remains a possible sense in which the conservation of genomic variation goes a long way to achieve conservation of all biodiversity: if it were established that speciation (as the epitome of a lineage diversification process) completely depends on genomic change, then we would have a good argument. The issue concerns the origin of biodiversity: if it turns out that genomic change is central, then some diluted form of global genetic reductionism might be rescued in the face of phenotypic plasticity (perhaps the variation produced through plasticity would be ineffectual *per se* for speciation; see West-Eberhard 2003 for an opposite argument).

As already argued above, the idea that a peculiar genome characterises a biological species remains common in biology (despite being generally rejected in philosophy of biology). This seemingly clear link between genetic properties and species partly explains why the latter are considered an element of the trinity. After all, species are the repositories of all the genomic and phenotypic variation among its constituent organisms, where this variation is the raw material on which speciation processes work. According to Mayr (1969), species are the most fundamental unit of biological organisation. Interestingly, Mayr argued that only sexually reproducing organisms form species and that the category is not applicable to many unicellular groups of organisms (e.g., bacteria), thus betraying a bias that still characterises conservation practice too. Mayr's (1969, p. 316) argument was that species serve a specific biological function because dividing the total genetic variability of nature into discrete packages prevents the production of "disharmonious incompatible gene combinations". Conservation efforts that target species could therefore be justified as we would save all the possible "harmonious genetic combinations".[7] However, even if we endorse the view that species are important units of biological organisation, this would not be enough to justify an exclusive focus on this biotic unit in conservation practice. One reason is that estimating biodiversity through species count is problematic.[8] For instance, species diversity would not account for diversity at other levels of the same hierarchy. The fact that there are more terrestrial than marine species does not translate into more diversity at the next hierarchical level; in fact, there are more marine than terrestrial phyla, i.e., diversity and disparity clash (Grosberg et al. 2012); hence, by conserving an equal number of marine and terrestrial species, we might not conserve equal marine and terrestrial biodiversity at the phylum level. Conversely, species diversity would not account for genetic diversity, that is, for diversity at another level of a different hierarchy (or even of the same hierarchy if the general taxonomic hierarchy proposed by Sarkar mixing genetic and taxonomic units is endorsed); hence, for instance, by choosing to conserve indiscriminately either species S1 and S2 of genus G because one of the two is functionally redundant (in the sense that they play an equivalent ecological role in the ecosystem), we might not be able to conserve equal biodiversity at the genetic level; the reason is that one of the species might harbour more genetic diversity (its gene pool might be larger); so, supposing chimps and bonobos play equivalent ecological roles in the ecosystems, conserving bonobos with presumably much smaller gene pools than chimps (Prado-Martinez et al. 2013) would amount to failing to conserve genomic diversity.

Similar arguments apply to exclusive focus on ecosystems as the unit of conservation. This means that it is clearly difficult to justify biodiversity fundamentalism.

---

[7] By adding the hypothesis that genes are the most important causes of phenogenesis, we end up with the strong hypothesis that by conserving the species' characteristic gene pool (i.e., an aggregate of genomes) we are also conserving the entirety of their possible phenotypic manifestations (that is, all protein and cell types as well as all supra-cellular organismal traits), i.e., all genetic and phenotypic biodiversity.

[8] See Borda-de-Água, Chap. 5 and Crupi, Chap. 6, in this volume on this issue.

A similar position has been argued for by Angermeier and Karr (1994, p. 691). They generalise the failure of biodiversity fundamentalism by also arguing that even a focus on a single hierarchy is bound to fail, as it would lead to ignore most biodiversity. Noss (1990, p. 357) has made this point quite succinctly by arguing that *"No single level of organization (e.g., gene, population, community) is fundamental ...."*. Of course, the idea of taking into account 3 units of different hierarchies instead of one unit is exactly tailored to avoid such problems. But, as Sarkar (2002, p 138) has argued, "...even this catholic proposal falls afoul of the diversity of biological phenomena ...". Thus, Sarkar argues, even avoiding biodiversity fundamentalism in some of its two forms (either focusing exclusively on a hierarchy or on a unit) would not allow accounting for "endangered biological phenomena" that are in principle amenable to conservation, such as the synchronous flowering of particular bamboo species at a distance. Sarkar argues that in order to save this peculiar phenotypic outcome, conservation efforts should neither be directed to conserve the genome of the clumps of these bamboo species, nor even conserving the species; rather, what should be conserved are the environments in which this behaviour is expressed; only by also preserving the habitats and sites where these biological phenomena occur we would be able to conserve them. Note that this argument is analogous to the one proposed above concerning the conservation of developmental environments. Developmental environments are, like habitats and sites, entities belonging to the spatial compositional hierarchy, that is, a different hierarchy than that to which genes and populations belong. We conclude that for all these reasons there is no justification for focusing exclusively on one compositional hierarchy in conservation practice. As we have showed, at least one spatial and one taxonomic unit are concomitantly needed as conservation units in order to encompass all phenotypic biodiversity (e.g., protein conformations and developmental outcomes, genetic and phenotypic variants) and all biological phenomena (e.g., synchronous flowering of bamboo). The corollary of this conclusion is that no biodiversity unit can be the fundamental unit of conservation. Rather, a variety of units are needed to encompass all biodiversity. We shall now suggest that the limits of biodiversity fundamentalism (both in its hierarchy and unit variants) and of multi-unit approaches to conservation is arguably a symptom of a more general malaise concerning entity-based approaches to conservation practice. The fundamental question is thus whether a different kind of approach should be favoured. In particular, we ask whether there exists a rationale for the exclusive focus on entities.

12.2.2   Towards an Entity and Process-Based Approach to Conservation

As we argued in the first section, one limit of compositional hierarchies pertains to their lack of structural and functional information. The problem is thus whether conservation strategies can be devised in the absence of detailed knowledge concerning the structural properties and functional interactions between the entities constituting the compositional levels of the hierarchy. Structural hierarchies aim to represent the organisation (e.g., the topology or network of interactions) between the parts of the relevant entity-whole, while functional hierarchies map the processes governing the causal interactions between the various parts of the relevant entity-wholes. In this sense, a structural characterisation of, for instance, a cell is the topology of the network of interactions between its components parts. The structural characterisation is not merely a list of cellular components (it is not purely compositional), but it is an organised list whereby their interactions are identified. A structural characterisation is more informative than a compositional one, but is less informative than a functional one. From a functional point of view, a cell is literally an ecosystem whereby energy and matter acquired from the environment is processed internally in such a fashion as to manufacture its component parts (Luisi 2003). This means that a functional representation of a cell specifies the causal nature of the interactions between its sub-cellular components. Consider secondly that, given that functional hierarchies aim to represent the causal interactions between the elements of a hierarchy, they do not provide merely entity-based ontologies. For instance, a functional characterisation of the cell makes reference to the metabolic interactions between nutrients, constituent proteins and other macromolecules, organelles, membrane receptors etc. In this sense, it does not purely provide an entity-based ontology but also a process-based one. The upshot is that the genetic, taxonomic and ecological hierarchies for characterising biodiversity in terms of genetic organisation, taxonomic inclusiveness and ecological nestedness are, given their compositional ethos, insufficient to capture the structural and functional aspects of biodiversity (Franklin 1988). It is for this reason that compositional hierarchies should be complemented with structural and functional hierarchies, as suggested by Noss (1990, p. 359). As soon as we look at functional hierarchies, we grasp that the focus is also on processes, not merely on entities.

The crucial question is whether knowledge of functional interactions and process is necessary in order to provide a satisfactory characterisation of biodiversity and especially of the units of conservation. Consider functional interactions first. Many species are involved in complex biological relationships such as predation and pollination. Compositional hierarchies provide information concerning the relata (i.e., the entities involved in a relation) of such interactions, but this information is oblivious to process. Pollination is an ecological function that can be realised in multifarious ways by a variety of species of insects, birds, bats, snails etc. on the one hand and flowering plants on the other. Perhaps some species play a fundamental ecological role in the pollination process (as keystone species, Sarkar 2005, p. 15) and our conservation efforts should be focused on these.[9] It is therefore clear that knowledge

---

[9] The concept of keystone species can be characterised in terms of ecological centrality (when a species has many functional relationships with many different species). This characterisation, however, seems to imply a lack of specialisation on the part of the species. For instance, pollination seems to be realised in large part by highly specialised species (both plants with very few pollinators and animals pollinating very few plants) which do not have, as a consequence, many functional relationships with many different species. We would argue that the ecological centrality of a keystone species depends on its specialised functional role: a keystone species would thus be one

of this ecological role might inform conservation efforts. However, knowledge of this kind is clearly not provided by compositional hierarchies. Consider processes now. Generally speaking, two types of processes governing the behaviour of the entities identified by compositional hierarchies can be identified. First of all, those leading to the differentiation of parts. Secondly, those that, given the differentiated parts, govern their combination (i.e., combinogenesis). All natural sciences are somehow concerned with understanding the nature of the processes of part differentiation and those governing their combination. Biology certainly strives to understand differentiation and combinogenetic processes: biology is both about differentiation of part-entities (e.g., production of genetic variants, new species etc.) and about the combinogenesis of whole-entities (i.e., the emergence of new biological individuals). For instance, what are the processes that govern allelic, population and species differentiation? Theories of genomic change and speciation are part and parcel of biology of course. And what are the processes that govern genome and ecosystem formation? Equally, theories concerning genome evolution and ecology are part and parcel of biology. An entity-based approach to biodiversity is thus parasitic on biological theories concerning, among others, genomic and phenotypic change as well as biological and ecological theories concerning, among others, genome evolution, phenotypic evolution, speciation and ecosystem stability, where all these theories make a reference to processes (e.g., mutation, phenotypic plasticity, predation) impinging on a variety of biological entities belonging to various levels of various compositional hierarchies. Thus, given that reference to such processes remains invisible in compositional hierarchies, they seem by their own nature epistemologically deficient. This point is particularly relevant because it influences the characterisation of the units of conservation. Does a focus on units of biodiversity, which are biological entities, make sense without a complementary focus on their maintenance and generative processes?

12.3 Does a Process-Based Approach to Biodiversity Make Sense?

We have argued so far that an entity-based approach ignores the functional relations between the elements of the hierarchy. In a nutshell, it ignores the influence of processes of differentiation and combination of parts. In conservation science, a process-based approach would shift the focus on the processes that *originate and maintain* biodiversity. As we shall relate, the shift from entities to process has been advocated by many conservation practitioners. The argument that we shall propose does not advocate a switch to exclusive focus on process. More reasonably, we suggest that a process-based approach should integrate an entity-based one (Faith

---

performing (almost) exclusively a particular function (e.g., pollination) for other species. See Sect. 12.3 for a clear example of keystone species (Morris et al. 2012). Thanks to Alessandro Minelli for drawing attention to this putative tension.

2016). After all, what could it mean to conserve a process? Not much. As we already argued in Sect. 12.1, if an exclusive entity-based approach to biology does not make sense, even an exclusive process-based approach does not. The reason is obvious: processes are important because they create and maintain new entities, new units of potential conservation. Thus, to use the pollination example again, the issue is not whether we should either choose the relata (e.g., the populations) or the relationship as units of conservation. We cannot think of any other sensible way of conserving relationships and processes than by conserving their relata and their actors (i.e., the entities involved in the process). As we shall explain below, the shift to process is most prominently a shift in the ways in which we characterise the units of conservation. Particularly important in the present context is Ryder's (1986) proposal to characterise conservation units as evolutionary significant units (i.e., ESUs), that is, as populations of organisms that, for historical and evolutionary reasons, play peculiar causal roles in the processes targets of conservation. From a process-based perspective, the ultimate focus of conservation practice is on entities such as ESUs (Moritz 1999, p. 223).[10] Relatedly, an important issue about the characterisation of a process-based approach concerns the kind of processes that should be taken into account. Noss (1990) considers as potential targets conservation processes that are partially abiotic such as energy cycles. Noss's is an interesting suggestion. However, it should be highlighted that, again, the focus is, ultimately, inevitably on the entities that play specific causal roles in processes. For instance, in marine environments some bacteria seem to play the role of keystone species as they might exclusively perform some specific function. For example, a limited number of bacteria (e.g., of the genus *Alteromonas)* seem to process hydrogen peroxide in the ocean, performing a crucial metabolic function that benefits the incredibly large communities of the cyanobacterium *Prochlorococcus* (Morris et al. 2012). Without these bacteria, the ecosystem would probably suffer. Conservation efforts could thus be directed to save this important geochemical process, but inevitably such efforts would focus on preserving the important ecological function that *Alteromonas* bacteria play. We thus suggest that the focus should be on the processes that govern what we called entity differentiation and combinogenesis in Sect. 12.2.2, that is, most prominently the ecological and evolutionary processes that cause the origin of ESUs.

Many inter-linked themes prominent in the conservation literature explain the shift towards a complementary entity-process-based approach to conservation. This conceptual shift finds its theoretical support in the deeper integration with the evolutionary sciences and with ecology. Most generally, Norton (2001) argues that conservation science has experienced a transition from a static to a dynamic view focused on evolving systems and ecosystem processes. This interpretive hypothesis is probably supported by a shift in the characterisation of the units of conservation from static entities – e.g., species characterised essentialistically in terms of species-specific genetic and phenotypic features – to historical ones with peculiar historical

---

[10] Sarkar (2002, note 15, p. 152) has argued that focus on process is aimed to conservation of biological "integrity" rather than biodiversity. However, if the focus of a process-based approach is on entities such as ESUs, it is clearly committed to biodiversity conservation.

and evolutionary capacities. This transition has been nurtured by the dissatisfaction with prominent species approaches to conservation aimed at the maximisation of number of species saved per spatial area which are, by definition, fundamentalist and entity-biased. Ultimately, the idea is that the conservation focus should be put on the evolutionary and ecological causes of biodiversity and the preservation of process rather than on their causal effects and on the preservation of pattern. Particularly important are the attempts to identify centres of evolutionary diversity with the aim of maximising evolutionary heritage on one hand (a consequence of the integration of phylogenetic analyses with conservation biology) and the focus on the evolutionary (e.g., genetic, cf. Frankel 1974) potential of populations and historical lineages. Smith et al. (1993) argue that knowledge of the ecological and evolutionary mechanisms generating genetic diversity and of the isolating mechanisms of speciation must be part and parcel of conservation practice. Conservation practices that focus on protecting species-rich sites are doomed to fail for reasons that parallel those for which the counting-species approach did. First, such focus does not necessarily provide information on the frequency of rare species, which might not occur in areas of highest species diversity (Smith et al. 1993, p. 164). Secondly, it does not necessarily provide any information on the functional role of species and on the nature of the community dynamics of the relevant ecosystems (Smith et al. 1993, p. 165). Thirdly, it neither necessarily identifies regions with peculiar evolutionary history nor identifies lineages that are phylogenetically unique (ibid.). For all these reasons, Smith et al. propose an approach to conservation that integrates ecological and molecular information. Related to the third point above, Mace et al. (2003) have suggested that, rather than directing conservation efforts to save species, these should be directed to saving independent branches of the tree of life, that is, distinctive lineages with a long and unique evolutionary history. The rationale for this conservation strategy is that phylogenetic information permits to distinguish "cradles" of diversity from "museums". A process-based approach informed by phylogenetic information (and hence by knowledge about evolutionary history) identifies as priority conservation taxa those that display a unique evolutionary history instead of focusing efforts on conserving patterns of species richness (Mace et al. 2003, p. 1709). Along the same lines of integration of molecular data, Moritz (1999) has proposed to address conservation problems by focusing on the maintenance and restoration of those ecological and evolutionary processes that can recreate adaptive phenotypes. In order to conserve such processes, we should aim to conserve their "effectors", i.e., the ESUs or populations with evolutionary potential in which they play causal roles. Moritz argues that molecular studies are particularly important to infer evolutionary history. Molecular information will give us details about the evolutionary relationships between the populations of conservation focus to the extent that, for instance, "… translocations among populations that historically exchanged genes would be considered, whereas human-mediated mixing of historically isolated gene pools would be discouraged." (Moritz 1999, p. 223). This approach aims to conserve ESUs through the restoration of connectivity between isolated populations in anthropogenically fragmented ecosystems and the destruction of "genetic ghettos" (Moritz 1999, p. 224). In synthesis, a process-based approach to conservation might be seen as proposing an integration of varieties of ecological and evolutionary information with the aim of identifying relevant ESUs. One general characterisation of ESUs that can be extrapolated from the conservation literature reviewed so far refers to populations of organisms possessing a property of conservation interest, such as a peculiar history (i.e., being a distinctive lineage) and a crucial functional role in ecosystem welfare (i.e., being a keystone species). Even though preservation seems to be, by definition, the aim of conservation biology, it is interesting to observe that Smith et al. (1993, p. 164) have argued that the aim of conservation science is "...to promote and preserve natural dynamics." What could promotion amount to? A promotion (rather than preservation) characterisation of ESUs might refer to properties of populations such as the ability to cope with environmental stress (i.e., adaptability) or an enhanced capacity to diversify into lineages with distinctive genetic and phenotypic features. In the latter two cases, it might be said that the population ESU displays "evolutionary potential" (Casetta and Marques da Silva 2015), a property that might depend either on possessing particular genomic properties or on its tendency to respond to environmental change purely phenotypically, where such properties might be important for populations' adaptability and diversification.[11] In the following section we shall focus on populations that display evolutionary potential in the latter sense. The hypothesis we would like to test is whether plastic populations of a species might be considered ESUs amenable to conservation. In particular, we would like to show that plastic subpopulations that have enhanced evolutionary potential vis a vis non-plastic subpopulations make them amenable to ESU status.

12.4   Can Phenotypic Plasticity Confer Evolutionary Potential?

In this section we shall thus focus on a particularly evolutionary process, i.e., phenotypic plasticity (Fitzpatrick 2012; Forsman 2015; Miner et al. 2005; Valladares et al. 2014; West-Eberhard 2003). By plasticity we refer to the ability of the organism to react to environmental inputs with an appropriate phenotypic change during embryogenesis (developmental plasticity) and further developmental stages (phenotypic plasticity). Two main types of plasticity exist: reaction norms and polyphenisms. In reaction norms the genome allows a continuous range of potential phenotypes. On the other hand, polyphenisms are discontinuous (either/or) phenotypes elicited by the environment. The essence of plasticity is that the genome does not wholly dictate the nature of the phenotypic outcome. It is reasonably straightforward to intuit about selective advantages to phenotypic plasticity: where there exist different or varying environmental conditions that are experienced, either a) by different individuals across a population, or b) by the same/each individual through its

---

[11] See Minelli, Chap. 11, this volume, for the relationship between evolutionary potential and evolvability.

lifetime, a unique phenotype (narrow reaction norm) would be less fit than plastic responses. This is, of course, provided that an appropriate phenotype can be expressed, either sensitive to environmental conditions or genetic (West-Eberhard 1986). An example of a genetic switch is the X-Y sex determination system in mammals. Some species of buttercups (e.g., *Ranunculus flammula*) exemplify polyphenism through environmental sensitivity: they develop one of two distinct leaf types, depending on whether underwater or on land (Cook and Johnson 1968). A particularly advanced form of environmental sensitivity – potentially producing continuous phenotypic responses – is learning: the capacity to *change* behaviour in particular situations, according to past life experiences (Staddon 1983).

While these phenotypic flexibilities are interesting in their own right, and indeed potential benefits of plasticity are easy to identify (notwithstanding discussion regarding what those benefits trade off against), could there be a deeper evolutionary issue here? Could phenotypic plasticity not only have proximate effects, but also impact the course of evolution? The understanding that traits produced through plasticity are not heritable goes as far back as the nineteenth century with August Weismann's experiments showing a soma/germ-line separation. And the hypothesis now commonly known as Lamarckian evolution, that traits acquired during lifetime would be passed on to further generations – e.g., the strong biceps of a blacksmith – is not considered compatible with genetic inheritance (discounting epigenetic inheritance). But there is an intriguing suggestion that phenotypic changes could influence selection in an evolving population, and thus indirectly lead to genetic encoding of formerly acquired traits (Baldwin 1896; Osborne 1896; West-Eberhard 2003). The basic notion is that the relatively rapid exploration of phenotype space via plastic response can introduce a selective gradient towards genetic specification of that phenotype, and thus the slower genetic variation can be "guided" by lifetime exploration (Hinton and Nowlan 1987). The selective landscape experienced by a plastic population is modulated by that plasticity, in comparison to the landscape experienced by non-plastic populations. But the modulation to fitness of specific genotypes does not require that the phenotypic traits discovered are heritable, i.e., it occurs without so-called "Lamarckian" inheritance. This process has become known as the Baldwin effect (a term coined by Simpson 1953). The effect depends on the existence of phenotypic plasticity (Bradshaw 1965) having already evolved but this in itself can only facilitate the first of two phases: selection among the various phenotypes expressed for those most appropriate to the present environmental conditions. The second phase, genetic assimilation, is not a necessary consequence of the existence of plasticity, nor does it depend on a reduction in the level of plasticity.[12] The Baldwin effect has been the subject of a plethora of computational studies (see e.g., Turney et al. 1996; Paenke et al. 2009; Sznajder et al. 2012), following the seminal work of Hinton and Nowlan (1987). Almost all of these works considered evolution in single-peaked fitness landscapes; but in Mills and Watson (2006) we showed that, via a Baldwinian process, a learning population is able to cross a fit-

---

[12] Mills and Watson 2005 further discuss how canalisation, although often implicated in studies on the Baldwin effect, is not actually a necessary mechanism for the effect.

ness valley. Here we use the same model to illustrate various scenarios, including that learning is able to repeatedly guide genetic evolution in a variable environment.

12.4.1   A Model of Plasticity

We model a population of individuals each with a string of n binary variables to represent their genotype, which specifies the phenotypes that the individual will express throughout its lifetime, through a trivial (but non-deterministic) genotype–phenotype (G-P) mapping. Specifically, for each lifetime trial i, the phenotype p_i is based on the genotype with mutation-like variation applied at a rate of μ_L, independently applied at each locus. The phenotypes from the T trials are independent from each other, and can be thought of as a cloud of points surrounding the genotypically-specified location. The individuals are bestowed with a simple capacity to learn, which is facilitated through the way that fitness is calculated: during each lifetime trial, a learning individual recalls the best solution found so far, whether it is the newest phenotypic strategy, or whether it was found long ago (see Hinton and Nowlan 1987). At the end of each generation, the individuals are selected in proportion to their fitness,[13] and reproduce asexually. During reproduction, point mutation is applied to each gene, i.e., each gene is transmitted to the offspring with a probability of $1\text{-}\mu_G$, otherwise with probability μ_G a new random allele is drawn (note that this model does not rule out the possibility of multiple mutations but that they are uncorrelated when they occur). The population size m is constant through time. In this model there is no way for an individual to perform less lifetime exploration, i.e., there is no mechanism for canalisation (Waddington 1953). This simplification is not meant to imply that there would never be a selective advantage to such a reduction, but rather to keep the spotlight on the consequences of plasticity.

Simulation Experiment 1 We consider the evolution of a population on a simple and abstract fitness landscape, where there are two rare phenotypes p_1 and p_2 that receive high fitness and all other phenotypes are equally bad. Here, our main question is to investigate whether the form of phenotypic learning in this model is sufficient for the population to evolve across the fitness valley between the two peaks. Accordingly, the first peak/phenotype confers high fitness ($f(p_1) = H$) and the second peak confers lower fitness ($f(p_2) = L$). The environment remains like this for s generations, after which the quality of peaks switches, such that $f(p_1) = L$ and $f(p_2) = H$.

Parameters used in this experiment: $H = 100$, $L = 10$, $f(p|p \neq p_1, p \neq p_2) = 1$, $n = 16$ genes. The separation d of the two peaks is 5 bits, and the switching interval s is 50

---

[13] Since all individuals experience the same number of learning trials before selection, it could be seen as selection only occurring on adult organisms; however, the model confers benefits to successful learning earlier in the lifetime, even though we do not explicitly include phenomena such as probabilistic death without reproduction.

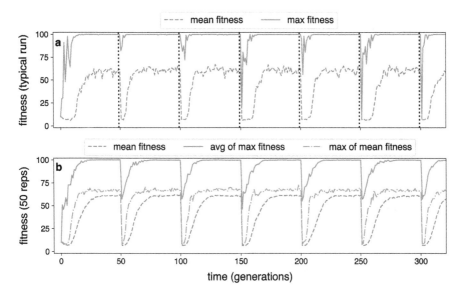

Fig. 12.1 Fitness measures of populations evolving in a switching environment. (**a**) one example run; (**b**) averages over 50 replicates. The switches in environment are marked by the dotted vertical lines

generations. We set the mutation rate μ_G at 1/20 and the lifetime variation rate μ_L at $2/n$, the number of trials per individual $T = 256$, and the population size m at 200.

To see what is happening in the population, we can observe the fitness over time (Fig. 12.1). Initially all organisms of the population possess the genotype specifying phenotype p_2, and within only a few generations some individuals in the population find a high-fitness phenotype, p_1. Any mutation that brings the genotype closer to directly specifying p_1 will be favoured since discovering the phenotype earlier in the lifetime results in higher fitness. Accordingly, such high-fitness genotypes propagate through the population, as is reflected in the rise in mean fitness. After each switch in the environment (dashed vertical lines), we see a sharp drop in fitness, reflecting the fact that the population was adapted to a previous challenge. However, phenotypic plasticity enables individuals to rapidly re-discover p_2, which is now the highest-fitness phenotype in the environment.

Simulation Experiment 2 Rather than fixing the rate of environmental switching, here we leave this parameter s open; and to ascertain the capacity of a plastic population to cope with such environmental change we run simulation experiments for various different values of T.

From the results in Fig. 12.2 we see two different trends: (1) populations experiencing a large number of trials T can achieve high fitness, provided the environment does not change too rapidly. When the interval is very short there is insufficient time for the population to find the high peak and assimilate it genetically. Note however that the high peak is found phenotypically by some fraction of the population: with-

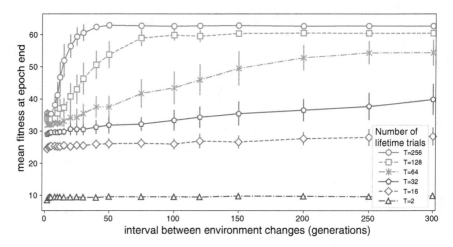

Fig. 12.2 Fitness of the population in the final generation before a switch. Data show the mean and standard deviation of mean fitness, across 40 replicates

out doing so, the mean fitness could not be greater than $f(L) = 10$. (2) The smaller the value of T, the less capacity the population has to adapt to the new challenge presented by the switched environment. At its extreme, with $T = 2$, the population wholly fails to adapt under any environmental switching rate tested.

The dynamics of a non-plastic population are qualitatively different. As the mutation model permits multiple loci to change simultaneously during reproduction, a multi-locus valley could, in principle, be crossed. However, the expected time this takes grows exponentially with the number of genes that must change at once. For the 5-bit valley and $\mu_G = 1/20$ (as used above), it takes a mean of over 75,000 generations, and even though a higher rate would reduce this, even an optimal rate of $\mu_G = 5/16$ takes a mean of 340 generations (mutation rates that are so high introduce difficulties in terms of drift and retaining high-fitness peaks even if/when discovered, besides severe penalties to average fitness). Importantly, these generation times are very high in comparison to the environmental switching frequencies that a learning population is able to thrive in.

The above model and experiments illustrate how one variety of phenotypic plasticity can enable a population to rapidly evolve across fitness valleys, a pattern of evolutionary change that cannot be experienced by non-plastic populations. On the flip-side, the benefits to the plastic population are lessened under more and more constant environments. At some point any benefits would be outweighed by the costs of learning (e.g., energetic cost of memory, risks). Although such aspects are omitted from the model here, and thus trade-offs are not directly visible in the results, the logic is straightforward: if genetic adaptation alone is sufficient in some stable environment, we should not expect to see plasticity playing any significant role.

Besides these two examples of environments, what other responses to changing environments might we expect to see in evolution? In the extreme case where many novel challenges appear within one lifetime, there may be plastic responses that do not become assimilated into the genome. In the absence of any regularity in those challenges, if the new challenges in one's lifetime are unrelated to the challenges faced by their ancestors, plasticity may be favourable but it is hard to see how any specific genetic adaptation would arise. If changes occur over a few generations, a Baldwinian-type interaction between plasticity and genetic evolution leading to genetic assimilation may result. Alternatively, if environmental changes are particularly repetitive, we may expect to see polyphenic/polymorphic genotypes and environmentally-sensitive switching (see West-Eberhard 1986, 2003) as mentioned above. If the environmental changes are strongly structured, we may additionally see modular architectures evolve in the genotypes (Parter et al. 2008) that are able to more quickly adapt to new challenges (Watson et al. 2014).

12.5   Conclusion

Our argument has been that focusing solely on entities, be they genes, species or ecosystems, is inherently problematic for conservation practice. We first argued that what we called *biodiversity fundamentalism* is untenable. It is both untenable as a thesis concerning the exclusive focus on one compositional hierarchy and as a thesis concerning the existence of a fundamental unit of conservation. Secondly, we argued that the genetic, taxonomic and ecological hierarchies for characterising biodiversity in terms of genetic organisation, taxonomic inclusiveness and ecological nestedness are, given their compositional ethos, insufficient to capture the functional dimension of biodiversity, particularly the evolutionary processes that maintain and originate new biodiversity units. Thirdly, we have proposed a *complementary entity-and-process-based approach to conservation practice*. Within this context, we distinguished between two types of important properties that evolutionarily significant units (i.e., ESUs) of conservation interest might exhibit: those amenable to conservation because they *preserve* natural dynamics (e.g., being a distinctive lineage) and those that *promote* them (e.g., being a population with a greater capacity for adaptation to change or stress). We focused on the latter because conservation strategies are aimed to identify not only "museums" but also "cradles" of biodiversity. Given this background, the hypothesis we wanted to test is whether plastic populations of a species might be considered ESUs with relevance for conservation. In particular, we wanted to show that plastic subpopulations that have enhanced evolutionary potential vis a vis non-plastic subpopulations make them amenable to ESU status. The model indeed shows that plasticity yields evolutionary potential, which is displayed in environments that switch in a few to a few tens of generations. Thus, populations with adaptation capacities available might possess an interesting property to consider when deciding on how to focus conservation efforts. Given that plastic populations might be important for species' adaptability and diversification, they might be considered ESUs potentially amenable to conservation. This vindicates, on the one hand, a process-based approach to biodiversity and, on the other, suggests the need to take into account the processes generating plasticity when considering conservation efforts.

Acknowledgements Many thanks to Sandro Minelli, Philippe Huneman and Elena Casetta for feedback. Davide Vecchi acknowledges the financial support of the Fundação para a Ciência e a Tecnologia (Grant N. SFRH/BPD/99879/2014 and BIODECON R&D Project. Grant PTDC/IVC-HFC/1817/2014). Rob Mills acknowledges support by UID/MULTI/04046/2013 centre grant from FCT, Portugal (to BioISI).

References

Angermeier, P. L., & Karr, J. R. (1994). Biological integrity versus biological diversity as policy directives: Protecting biotic resources. *Bioscience, 44*(10), 690–697.

Baldwin, J. M. (1896). A new factor in evolution. *American Naturalist, 30*, 441–451.

Bradshaw, A. D. (1965). Evolutionary significance of phenotypic plasticity in plants. *Advances in Genetics, 13*, 115–155.

Casetta, E., & Marques da Silva, J. (2015). Facing the big sixth: From prioritizing species to conserving biodiversity. In E. Serrelli & N. Gontier (Eds.), *Macroevolution: Explanation, interpretation and evidence* (pp. 377–403). Cham: Springer.

Cook, S. A., & Johnson, M. P. (1968). Adaptation to heterogeneous environments. I. variation in heterophylly in *ranunculus flammula* l. *Evolution, 22*(3), 496–516.

Dupré, J. (2012). *Processes of life: Essays in the philosophy of biology*. Oxford: Oxford University Press.

Faith, D. P. (2016). Biodiversity. In The Stanford encyclopedia of philosophy (Summer 2016 edition), ed. Edward N. Zalta. http://plato.stanford.edu/archives/sum2016/entries/biodiversity. Accessed 25 Sept 2018.

Fitzpatrick, B. M. (2012). Underappreciated consequences of phenotypic plasticity for ecological speciation. *International Journal of Ecology*. https://doi.org/10.1155/2012/256017.

Forsman, A. (2015). Rethinking phenotypic plasticity and its consequences for individuals, populations and species. *Heredity, 115*, 276–284.

Frankel, O. H. (1974). Genetic conservation: Our evolutionary responsibility. *Genetics, 78*, 53–65.

Franklin, J. F. (1988). Structural and functional diversity in temperate forests. In E. O. Wilson (Ed.), *Biodiversity* (pp. 166–175). Washington, DC: National Academy Press.

Grosberg, R. K., Vermeij, G. J., & Wainwright, P. C. (2012). Biodiversity in water and on land. *Current Biology, 22*(21), R900–R903.

Hennig, W. (1966). *Phylogenetic systematics*. Urbana: University of Illinois Press.

Hinton, G. E., & Nowlan, S. J. (1987). How learning can guide evolution. *Complex Systems, 1*(3), 495–502.

Luisi, P. (2003). Autopoiesis: A review and a reappraisal. *Naturwissenschaften, 90*, 49–59.

Mace, G. M., Gittleman, J. L., & Purvis, A. (2003). Preserving the tree of life. *Science, 300*(5626), 1707–1709.

Mayr, E. (1969). The biological meaning of species. *Biological Journal of the Linnean Society, 1*, 311–320. https://doi.org/10.1111/j.1095-8312.1969.tb00123.x.

Mayr, E. (1970). *Populations, species, and evolution*. Cambridge: Harvard University Press.

Mills, R., & Watson, R. A. (2005). Genetic assimilation and canalisation in the Baldwin effect. In M. S. Capcarrère et al. (Eds.), *Advances in artificial life* (pp. 353–362). Berlin: Springer.

Mills, R., & Watson, R. A. (2006). On crossing fitness valleys with the Baldwin effect. In L. M. Rocha (Ed.), *Proceedings of the tenth international conference on the simulation and synthesis of living systems* (pp. 493–499). Cambridge: MIT Press.

Miner, B. G., Sultan, S. E., Morgan, S. G., Padilla, D. K., & Relyea, R. A. (2005). Ecological consequences of phenotypic plasticity. *Trends in Ecology and Evolution, 20*(12), 685–692.

Moritz, C. (1999). Conservation units and translocations: Strategies for conserving evolutionary processes. *Hereditas, 130*(3), 217–228. https://doi.org/10.1111/j.1601-5223.1999.00217.x.

Morris, J. J., Lenski, R. E., & Zinser, E. R. (2012). The black queen hypothesis: Evolution of dependencies through adaptive gene loss. *MBio, 3*(2), e00036–e00012. https://doi.org/10.1128/mBio.00036-12.

Norton, B. G. (2001). Conservation biology and environmental values: Can there be a universal earth ethic? In C. Potvin et al. (Eds.), *Protecting biological diversity: Roles and responsibilities*. Montreal: McGill-Queen's University Press.

Noss, R. F. (1990). Indicators for monitoring biodiversity: A hierarchical approach. *Conservation Biology, 4*(4), 355–364.

Osborn, H. F. (1896). Oytogenic and phylogenic variation. *Science, 4*(100), 786–789. https://doi.org/10.1126/science.4.100.786

Paenke, I., Kawecki, T. J., & Sendhoff, B. (2009). The influence of learning on evolution: A mathematical framework. *Artificial Life, 15*(2), 227–245.

Parter, M., Kashtan, N., & Alon, U. (2008). Facilitated variation: How evolution learns from past environments to generalize to new environments. *PLoS Computational Biology, 4*(11), e1000206.

Prado-Martinez, J., Sudmant, P. H., Kidd, J. M., Li, H., Kelley, J. L., Lorente-Galdos, B., Veeramah, K. R., Woerner, A. E., O'connor, T. D., Santpere, G., & Cagan, A. (2013). Great ape genetic diversity and population history. *Nature, 499*(7459), 471–475.

Rosenberg, A. (1997). Reductionism redux: Computing the embryo. *Biology and Philosophy, 12*, 445–470.

Ryder, O. A. (1986). Species conservation and systematics: The Dilemma of subspecies. *Trends in Ecology and Evolution, 1*, 9–10.

Sarkar, S. (2002). Defining "biodiversity"; assessing biodiversity. *The Monist, 85*(1), 131155.

Sarkar, S. (2005). *Biodiversity and environmental philosophy: An introduction* (Cambridge studies in philosophy and biology). Cambridge: Cambridge University Press.

Schaffer, J. (2003). Is there a fundamental level? *Noûs, 37*, 498–517.

Simpson, G. G. (1953). The Baldwin effect. *Evolution, 7*(2), 110–117.

Smith, T. B., Bruford, M. W., & Wayne, R. K. (1993). The preservation of process: The missing element of conservation programs. *Biodiversity Letters, 1*(6), 164–167.

Staddon, J. E. R. (1983). *Adaptive behavior and learning*. Cambridge: Cambridge University Press. (Internet edition 2003).

Sznajder, B., Sabelis, M. W., & Egas, M. (2012). How adaptive learning affects evolution: Reviewing theory on the Baldwin effect. *Evolutionary Biology, 39*(3), 301–310.

Turney, P., Whitley, D., & Anderson, R. (1996). Evolution, learning, and instinct: 100 years of the Baldwin effect. *Evolutionary Computation, 4*(3), iv–viii. https://doi.org/10.1162/evco.1996.4.3.iv.

Valladares, F., Matesanz, S., Guilhaumon, F., Araújo, M. B., Balaguer, L., Benito-Garzón, M., Cornwell, W., et al. (2014). The effects of phenotypic plasticity and local adaptation on forecasts of species range shifts under climate change. *Ecology Letters, 17*, 1351–1364. https://doi.org/10.1111/ele.12348.

Waddington, C. H. (1953). Genetic assimilation of an acquired character. *Evolution, 4*, 118–126.

Watson, R. A., Wagner, G. P., Pavlicev, M., Weinreich, D. M., & Mills, R. (2014). The evolution of phenotypic correlations and "developmental memory". *Evolution, 68*(4), 1124–1138.

West-Eberhard, M. J. (1986). Alternative adaptations, speciation, and phylogeny A review. *Proceedings of the National Academy of Sciences, 83*(5), 1388–1392.

West-Eberhard, M. J. (2003). *Developmental plasticity and evolution*. Oxford: Oxford university press.

Wimsatt, W. C. (2007). *Re-engineering philosophy for limited beings: Piecewise approximations to reality*. Cambridge: Harvard University Press.

Wolpert, L. (1994). Do we understand development? *Science, 266*, 571–572.

13

Biodiversity as Explanans and Explanandum

Philippe Huneman

Abstract Biodiversity is arguably a major topic in ecology. Some of the key questions of the discipline are: why are species distributed the way they are, in a given area, or across areas? Or: why are there so many animals (as G. Evelyn Hutchinson asked in a famous paper)? It appears as what is supposed to be explained, namely an *explanandum* of ecology. Various families of theories have been proposed, which are nowadays mostly distinguished according to the role they confer to competition and the competitive exclusion principle. *Niche* theories, where the difference between "fundamental" and "realised" niches (Hutchinson GE, Am Nat 93:145–159, 1959) through competitive exclusion explains species distributions, contrast with *neutral* theories, where an assumption of fitness equivalence, species abundance distributions are explained by stochastic models, inspired by (Hubbell SP, The unified neutral theory of biodiversity and biogeography. Princeton University Press, Princeton, 2001).

Yet, while an important part of community ecology and biogeography understands biodiversity as an *explanandum*, in other areas of ecology the concept of biodiversity rather plays the role of the *explanans*. This is manifest in the long lasting stability-diversity debate, where the key question has been: how does diversity beget stability? Thus explanatory reversibility of the biodiversity concept in ecology may prevent biodiversity from being a unifying object for ecology.

In this chapter, I will describe such reversible explanatory status of biodiversity in various ecological fields (biogeography, functional ecology, community ecology). After having considered diversity as an *explanandum*, and then as an *explanans*, I will show that the concepts of biodiversity that are used in each of these symmetrical explanatory projects are not identical nor even equivalent. Using an approach to the concept of biodiversity in terms of "conceptual space", I will finally argue that the lack of unity of a biodiversity concept able to function identically as *explanans* and *explanandum* underlies the structural disunity of ecology that has been pointed out by some historians and philosophers.

P. Huneman (✉)
Institut d'Histoire et de Philosophie des Sciences et des Techniques, CNRS/Université Paris I Panthéon Sorbonne, Paris, France

Keywords Explanation · Species richness · Functional diversity · Phylogenetic diversity · Modern Synthesis · Neutral theory · Niche · Stability

13.1 Introduction

Amongst the questions that theoretical ecologists have been debating for decades one finds: why are species distributed the way they are, in a given area, or across areas? How is biodiversity related to areas? Why are there so many species in tropical regions? In general, why are there so many animals (as Hutchinson asked in a famous paper)? Is the amount of species currently decreasing and at what tempo? Why are so many species getting extinct in some environments now? Those questions have to do with what we have been calling, since Walter G. Rosen coined the word in the 80s (Takacs 1996) and Wilson (1988, 1992) popularized it, "biodiversity".

However, there are many ways of measuring biodiversity, tracking its progress or, more realistically, its erosion: different measurement methods defined by different indexes, such as Shannon index, Simpson index, etc. (Gosselin 2014; Noss 1990), as well as various ways of capturing it in relation to the ecological scale, such as beta diversity, gamma diversity,[1] etc. Moreover, there are several concepts of biodiversity, some attributing species a privileged role and others including also genes, or ecosystems, as is attested in the definition of biodiversity used in international conventions, such as the Convention on Biological Diversity (1992): "'Biological diversity' means the variability among living organisms from all sources including inter alia, terrestrial, marine and other aquatic ecosystems, and the ecological complexes of which they are a part; this includes diversity within species, between species and of ecosystems". And even at the level of species diversity, species richness as the mere amount of species is often considered too rough a biodiversity concept. In order to design robust diversity indices, ecologists or conservation biologists often add species evenness, and then consider the width of diversity, named as disparity – some wanting also to integrate the consideration of abundances within the concept of diversity (Blandin 2014). In addition to mere species counting, however, some dimension of species similarity sometimes ought to be included in the concept of diversity: mitigating species diversity by functional or phylogenetic similarity results in the concepts of phylogenetic diversity or functional diversity, whose use is especially required in ecophylogenetics (Mouquet et al. 2012) for the former and in functional ecology for the latter.

---

[1] Those terms were introduced by Whittaker to capture aspects of the local and regional distributions of diversity. *Alpha diversity* refers to species diversity on sites or habitats at a local scale as well as to the ratio of local to regional diversity, *beta diversity* compares the species diversity between ecosystems or across environmental gradients; *gamma diversity* is the total diversity in a landscape and therefore the compound of the former two.

Thus, while it was tempting in the beginning to consider biodiversity as a key question and a key *explanandum* of ecology, the diversity of biodiversity prevents us from straightforwardly claiming this. It may be argued, in turn, that this diversity seems to echo a lack of unity that affects ecology itself. It has indeed often been complained in ecology that the field lacks the unity that characterises the sister field of evolutionary biology. In 1989, Hagen already saw ecology as affected by a deep cleavage between a holological perspective and a mereological perspective, the latter using a demographic approach to ecosystems and communities while the former relies on a systemic view of the ecological objects, with or without appealing to evolutionary schemes of thought and natural selection (Hagen 1989). He concluded that this cleavage is essential to the discipline, and in turn allows ecology to explore a wide variety of objects and problems. More recently, Vellend (2016) has explicitly drawn a parallel between evolutionary biology and ecology and argued that ecology never had a unified framework similar to the one that structured evolutionary biology from the 50s onwards, and that allowed this science to flourish by providing researchers with common concepts, methods, key examples, key issues, and references.

Would it make sense to consider that the diversity of biodiversity is involved in the lack of unity of theoretical ecology? Or, more precisely, which disunity would be induced by this diversity, and is it unredeemable?

This will be the main question of the chapter. I will start by considering the issue of the long sought unity of ecology (13.2). Then I will explicate what I call the "explanatory reversibility" of biodiversity in ecology, namely its capacity to be *explanandum* and *explanans* in a science, as an essential feature of its theoretical role (13.3). Section 13.4 will consider more precisely the aspects of diversity as an *explanandum* of various ecological programmes, involving distinct explanatory schemes. Section 13.5 turns to diversity as an *explanans*, focusing on the relations between various kinds of stability and distinct notions of diversity, and characterizing the differences between such diversity and the way diversity is used in the explanatory programmes formerly described. In Sect. 13.6, I propose an account of the ecological notion of diversity in terms of a "conceptual space", in which various biodiversity concepts used in the varied explanatory strategies I described are specifically constructed. I use it in order to explicate the specific profile of the explanatory reversibility of diversity in ecology, and draw conclusions about the lack of unity in ecology and the epistemic status of the notion of diversity. The major argument developed there relies on the fact that the two explanatory projects concerning diversity target different "regions" of the total conceptual space of biodiversity so described.

13.2   The Unity of Ecology

It is often heard that ecology lacks unity – be it to complain about the missing unity (Vellend 2010), or to claim that it is a richness proper to this scientific discipline

(Hagen 1989). Inversely, "unifying principles" or theories have been constantly pursued for ecology (e.g. Margalef 1963; Hubbell 2001; Loreau 2010; Vellend 2016). Before considering the specific theoretical role of a biodiversity concept in ecology, and the possible unifying role it could play as an object or a pervasive concept, I will review the most general divides that seem to prevent such unity. After having listed some subfields, I will attempt at ordering this disunity by indicating the major lines of division (summarized in Fig. 13.1 below).

A quick glance at ecological subdisciplines shows the overall variety of questions and methods that characterizes the field. *Behavioural ecology* studies the traits ("behaviours") of organisms, hypothesized as adaptations to their (possibly social) environment; *community ecology* is about communities, i.e. sets of various species in the same region, considered from the viewpoint of the diversity and succession of species occurring within it. *Population ecology* mostly considers few species and focuses on the dynamics of the abundances of each of them given their major ecological interactions (predation or competition). *Biogeography* is interested in the distribution of species across higher scale dimensions, namely regions. *Functional ecology* considers the interactions between various species from the viewpoint of their net effect on the shared environment, especially by addressing networks of trophic relations and ascribing its species a role in the ecosystem (Loreau 2010). *Ecosystem ecology* as advocated by the Odum brothers (e.g. Odum 1953) develops such approach and uses schemes of thermodynamic thinking in considering ecosystems (i.e. communities plus their abiotic environment) under the perspective of semi-closed systems exchanging matter and energy with their environment (Hagen 1992). On the other hand, *Ecological genetics* initiated by E.B. Ford (Ford 1964) – a student of Fisher – considers the dynamics of population in various species from the standpoint of the changes of gene frequency within each species. Finally, *evolutionary ecology* (Roughgarden 1979) borrows tools from ecological genetics and approaches ecological patterns as results of evolutionary processes.

Fig. 13.1 The divides of ecology. Each thick line represents one of the four dimensions. The thin lines stand for the position of three historically important views in ecology: in black, Clements (1916); in red, Allee et al. (1949); in blue Nicholson and Bailey (1935) (Color figure online)

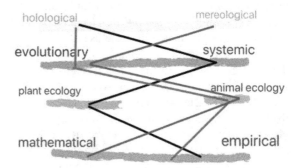

Broadly speaking, evolutionary biology and ecology have complex relationships (McIntosh 1986, pp. 256–263; Collins 1986; Harper 1967; Stearns 1982; Antonovics 1976; Huneman 2019). Besides its interest in explaining adaptation and evolution (phylogenetic patterns), evolutionary biology is interested in explaining the patterns of diversity and unity across diversity (i.e. homologies and analogies) that characterize extant and past taxa (as well as molecular patterns). And ecologists are generally interested, as I will argue more extensively in the following, in diversity within and across communities and ecosystems. Thus, both disciplines, at their own timescale, focus on the same *explanandum,* that is *diversity.*

Haeckel famously defined ecology as the "science of the struggle for existence", thus directly tying it to evolutionary biology to the extent that natural selection is seen by Darwinians as the key *explanans* and cause of evolution, adaptation and diversity. In principle, the emphasis on natural selection can be more or less strong in ecology, and this characterizes the whole field of ecological sciences: they are somehow ordered along a gradient which goes from evolutionary ecology to ecosystem ecology. At one extremity, evolutionary ecology adopts a very *evolutionary* viewpoint, considering ecosystems as the scene of competition, cooperation and mutualistic interactions, all occurring in evolutionary time and therefore being always dynamic. The other extremity of the continuum may be represented by most trends in ecosystem ecology, which adopts a very *systemic* viewpoint (sometimes akin to thermodynamics) insofar as ecosystems are open dissipative systems, more or less chaotic, dealt with in thermodynamic or statistical mechanics terms (Hagen 1992).

This divide is not the same as Hagen's distinction between *mereological* and *holological* perspectives on ecology mentioned before, since the holological view would accept an evolutionary understanding of ecology that takes communities or ecosystems as targets of selection. For instance, within a holological view echoing a Clementsian concept of community, an evolutionary parallel between communities and organisms, both being shaped by natural selection, is explicitly drawn in one of the major works on animal ecology in the mid twentieth century, namely the *Principles of animal ecology* written by Chicago ecologists Clyde Allee, Thomas Park, Orlando Park, Schmitt and Alfred Emerson, and praised in the *American Naturalist* by Dobhzansky. They say: "a community may be said to have a characteristic anatomy, an equally characteristic physiology and a characteristic heredity", therefore community is the "smallest [unit] that can be (…) selfsustained", and is precisely *"a resultant of ecological selection"* (Allee et al. 1949, 437).

Besides this gradient around the use of selection, which ranges across ecology, from the most systemic explanatory schemes (e.g. Odum's ecosystem ecology) to the most evolutionary understanding (e.g. Roughgarden's evolutionary ecology), ecology has been disciplinary cleaved between *plant ecology* and *animal ecology* since the 1900s. The two traditions were developed quite separately, starting respectively with the major advances of Warming (1909) and then Clements (1916) in plant ecology on the one hand, and the attempt at systematising animal ecology by Elton in the 1920s (e.g. Elton 1927), on the other hand.

Clements' idea of succession in communities, analogous to the development of organisms, was a key concept for much of plant ecology (Horn 1975; Lortie et al.

2004). Even though the individualistic concept of community put forth by Gleason in the 30s (Gleason 1939) won over much of ecology, one can still see a difference between plant and animal ecology to the extent that the attention paid to succession and assemblages has been more prevalent in plant than in animal ecology. And as noted by Harper (1967), it is harder to track the offspring of plants and determine their reproductive success, which partly explains why evolutionary perspectives were less favoured in plant ecology than in animal ecology.

In turn, animal ecologists have been massively worried about the question of the regulation of population size, which was probably the major controversial theme of ecology in the 50s, as can be noticed in the major gathering of evolutionary biologists and ecologists at the Cold Spring Harbour Symposium in 1957, devoted to population biology. Most of the talks – by Anderwartha, Birch, Lack, Chitty, Orians, etc. – were about population regulation in animal ecology. Much of this interest in regulation of population stemmed from a concern about pests. Charles Elton, a pioneer of invasion ecology, was the founder of the Bureau of Animal Populations in Oxford and one of its important tasks was pest control (see Chew (2011) on historical overview of invasion ecology, and Richardson (2011) on Elton's legacy in the field). Understanding the reasons of population regulation, population cycles and possible overpopulation was a crucial requisite for a successful control. One may argue that this context explains the difference between plant and animal ecology regarding the prevalence of the population regulation issue.[2]

Orthogonal to this divide between plant and animal ecology, there is an important tension between a more *empirically* oriented ecology and a mostly *mathematical* ecology (e.g. Schoener 1972). In a 1949 paper on population regulation (Solomon 1949). ME Solomon, British ecologist of the Bureau of Pest Control, noticed that ecologists are divided into two camps, one that starts from biology and generalizes, and one that builds mathematical models first and then tries to fit in the biological facts – e.g. Thomas Park's experiments on flour beetles, or perturbation experiments on populations (Smith 1952) vs. Lotka and Volterra's equations. This divide (see Kingsland 1995) still persists in various modes, as indicates the need recently felt by some theoretical ecologists to vindicate the use of mathematical theorizing (Servedio et al. 2014).

Regarding those five distinctions, each theoretical construction can be situated on each of the axes constituted by the gradient occurring between the poles of the distinction. In Fig. 13.1 I sketched the position of very influential works taken from distinct periods of the history of ecology (Allee et al.'s *Treatise* (1949), Nicholson and Bailey's model of host-parasite dynamics (1935), and Clements' plant ecology (1916)).

---

[2] Actually, Clements and Shelford (1939) intended to close the gap between plant and animal ecology, by applying a very general concept of community. They say: "the development of the science of ecology has been hindered in its organization and distorted in its growth by the separate development of plant ecology on the one hand and animal ecology on the other." (p.v) Ten years on and with a similar goal of systematizing ecological knowledge and providing basic principles, Allee et al. (1949), while acknowledging that principles of ecology should be general, still restrained to animal ecology for reasons of immaturity of the field.

However, in the face of these various divides within ecology, someone could argue that *biodiversity* defines an object of investigation that crosses frontiers between traditions, paradigms, and explanatory strategies. A major issue in ecology is indeed *coexistence* – why is it that certain various species coexist and others not? How can they do so? It is a question in both plant ecology and animal ecology, approached from mereological as well as holological perspectives, and through mathematical or more empirically oriented perspectives as well. Community ecology is openly concerned with explaining biodiversity patterns, and biogeography enquires about species-area laws, which are patterns about how biodiversity is scattered across various kinds of areas (MacArthur and Wilson 1967). However, even functional ecology gives a key role to diversity (Loreau 2010), at least under the mode of "functional diversity", namely the differences between species partitioned according to equivalence classes defined by ecological functional roles (producer, nutrient cycler, etc.) (Dussault and Bouchard 2017).

In addition, the emergence of the word 'biodiversity' in the 1980s could also indicate that there is an object proper to ecology here. In the following I shall focus on the explanatory logics of diversity in ecology, in order to assess (and eventually infirm) the hypothesis that biodiversity constitutes a shared object amongst various ecological theories and traditions, and that its concept could help define a unifying framework for ecology.

13.3   The Explanatory Reversibility of Diversity

Many ecologists' researches indeed focus on diversity. They range from very general questions about what causes diversity in general – Hutchinson asking "Why are there so many animals?" (Hutchinson 1961)–, to questions about the way diversity is distributed locally and regionally – species-area laws in biogeography, and the mathematical models explaining them in McArthur and Wilson's *Theory of Island Biogeography* (1964), species abundance distributions in community ecology, or the patterns of succession of plant species in communities (as illustrated in Clements' works), as well as questions of medium degree of generality about how it is possible that many species coexist *generaliter*. One could view all these questions as various modes of an overarching coexistence question: how is coexistence (amongst *diverse* species or organisms) possible and realised at various scales?

Besides *explaining* diversity under its various modes, ecology is concerned with biodiversity in another and very different way. A longstanding debate in ecology regards what has been labelled the "diversity-stability hypothesis" (Ives and Carpenter 2007; Pimm 1984). Simply put, it is the claim that diversity – especially species richness – begets stability (mostly in the form of the constancy of species abundances). The more species an ecosystem includes, the more stable it seems to be (namely, it contains the same species for a long time, with abundances fluctuating around a steady means). In this sense tropical forests, which are species rich ecosystems, have been providing examples of this pattern for many decades. The intuition of this fact was very robust, but its explanation has been overlooked for a long time.

However, in 1974 when Robert May started to investigate this hypothesis mathematically, by modelling networks of species and increasing the diversity value, it turned out that diversity does not beget stability but on the contrary prevents it (May 1974). Assuming that a system which is "stable only within a comparatively small domain of parameter space (…) may be called *dynamically fragile*", clearly "such a system will persist only for tightly circumscribed values of the environmental parameters" (May 1975). The result of May's models is that a "wide variety of mathematical models suggest that as a system becomes more complex, in the sense of more species and a more rich structure of interdependence, it becomes more dynamically fragile". (ib.) Researchers then tried to address this gap between these mathematical models and some data that tended to show a stability-friendly effect of diversity. The question of stability then became: what does explain the fact that some empirically attested diversity does not conform to May's mathematical models?

Ecological stability is actually a crucial issue for theoretical reasons. After Darwin's revolution, Linnaeus' explanations for stability (namely, each species fulfills a role in a well-balanced nature; see Pimm 1993) were no longer possible, and, inversely, the constancy of ecosystems constitutes a challenge if the world is an ever-changing Darwinian world led by competition. Ecological stability is also challenging for practical reasons, since understanding what makes ecosystems robust could allow us to manage and protect them. (In fact, almost since its beginnings scientific ecology has been concerned with the damages inflicted by human industry and agriculture to natural ecosystems and ultimately to the environments in which human societies live).

Diversity is therefore a two-faceted concept: it is a major *explanandum* for ecology under various guises, but when the question concerns the stability of ecosystems, diversity becomes an *explanans*. We witness here a major epistemological feature of evolutionary and ecological questions, namely the "explanatory reversibility" of key concepts. Some concepts may indeed be the *explanandum* in some contexts and the *explanans* in others, and this reversibility attests to their theoretical significance. In evolutionary biology, notions such as plasticity (Nicoglou 2015), robustness (Wagner 2005; Huneman 2018) or mutation rates display this epistemic feature, which was first recognised by Fisher in connection with some major properties of the genetic system (dominance, recessivity, etc.) that condition evolution and are at the same time a product of past evolution (Fisher 1932).

In the field of ecology, diversity constitutes one of the concepts whose epistemic profile displays such reversibility. In the following I will explore this reversibility in more detail, and examine the role it may play in the structure of ecology.

13.4   Diversity as an *Explanandum*: Conceptual and Historical Aspects of the Ecological Coexistence Issue

From the early times of ecology, diversity as an explanandum has been understood as a question of coexistence. I shall recapitulate this matter, and then consider a theoretical framework used to address it, namely the concept of ecological niche and its formulation by Hutchinson. I shall then turn to rival conceptions, mostly structured today around the idea of a "neutral theory". In each case, I will emphasize the aspects of the concept of biodiversity that are prominently addressed in each explanatory scheme.

The *coexistence* question may be arguably one of the key issues handled by ecologists since the early twentieth century. For decades, plant ecologists have embraced Clements' concept of community, which is slightly like an organism and displays a process of succession analogous to the development of an organism.[3] Clements and the animal ecologist Sheldon in *Bioecology* (1939) generalized this idea to plant and animal communities. Allee et al. (1949) major treatise on animal ecology took the concept of community on board – i.e. "the natural unit of organization in ecology" (437) – as well as the parallel between organisms and communities, since like Clements they consider communities to have a "metabolism". Their question here is about explaining the composition rules of an assemblage of species in a given community, and whether there are laws governing these species' procession.

However, in the 40s and 50s, the coexistence question seemed to be supplanted by a different issue, i.e. the explanation of *population regulation*: why does a population of a species generally fluctuate over a specific abundance, with regular cycles? From Elton (1927) to Hutchinson (1957) at least, the regulation issue was the other major problem for ecologists, especially animal ecologists – with, as mentioned above, a practical concern for invasions and pest control. To some extent, the regulation issue was more mathematically tractable than that of coexistence, as attested in the seminal models by Lotka and Volterra (Volterra 1926) and by Nicholson and Bailey (1935), which mostly deal with two or three species. Nicholson and Bailey explicitly acknowledged that handling many species would require very sharp mathematical skills (1935, 597).

Yet, when Hutchinson (1957) formulated his influential concept of niche as a hyperspace of environmental parameters in which a subspace of the hyperspace defines the viability conditions for a species, the coexistence question came again at the center of theoretical ecology.[4] At the time, such question was often traced back to an appeal to some form of group selection, as exemplified by Allee et al. (1949) animal ecology treatise and shared by many ecologists, as indicated above (see

---

[3] "Development is the basic process of ecology, as applicable to the habitat and community as to the individual and species." (Clements and Shelford 1939, 4)

[4] See Pocheville (2015) for a conceptual history of 'ecological niche' that relates Hutchinson to earlier views by Grinnell and Elton.

Mitman (1988) on these ideas of collectives and group selection). Coexistence in a community could be thought along the same lines as organismic integration, given that natural selection – individual in the latter case, collective in the former case – underlied both systems and their cohesiveness. David Lack's work on clutch size (Lack 1947, 1954) however progressively provided powerful arguments to think that individual natural selection, and not group selection, was the reason of population regulation, and a little bit later the idea of group selection met the devastating critique issued by Williams (1966). All this made the group selection approach to the coexistence question harder in principle. Hutchinson's idea of niche to some extent thereby set the frame for more fruitful approaches to various modes of the coexistence question.

More precisely, Hutchinson published his conception in the "Concluding remarks" to the 1957 Cold Spring Harbour Symposium, where prominent ecologists and evolutionary biologists debated population ecology and mostly the regulation issue.[5] The volume was a final landmark in the debate over competition-centered (inspired by Nicholson (1933) initial model of regulation by density-dependent factors, and mostly represented by Lack (1954)) and density-independence-centered explanations of population regulation that emphasized factors such as climate (Anderwartha and Birch 1954).[6] Hutchinson's view of the niche followed his assessment of the debate, which tried to fairly acknowledge some epistemic value in both positions – mainly Lack's view of density-dependent regulation by competition and Anderwartha and Birch (1954) view of regulation by density-independent factors.

This concept of niche was used by Hutchinson to make sense of the role of competition in the regulation process. But more importantly, it also allows a grasp on the *coexistence* issue. Here, what explains coexistence is indeed the fact that first, each species has a "fundamental niche", and second, that the portion of a fundamental niche shared by two species will be exclusively inhabited by the best competitor (Fig. 13.2). "Fundamental niches" once restricted by the process of competition – so, finally, natural selection – yield the "realised niches", which explain where a species will actually be found in the environment. In a classic study, Joseph Connell (1961) studied two species of barnacles, *Balanus balanoid* and *Chtamalus stemallus*, which have a stratified distribution along the coast of Scotland. The *Balanus* live on the border between see and rock, while the *Chtamalus* live just above it (Fig. 13.3a). *Balanus* cannot really live much higher because they cannot resist dessication during low tides. But if we take out the *Balanus*, the *Chtamalus* now appear to occupy also the space inhabited by the *Balanus*, in addition to their known territory (Fig. 13.3b). Thus, the fundamental niche of *Chtamalus* is the whole region of the rocks on which *Balanus* and *Chtamalus* live, but their realised niche is the territory where one finds them along with the *Balanus*, because the latter are a better competitor and wash *Chtamalhus* away from this portion of their fundamental

[5] On Hutchinson's work and influence on ecology see Slack (2010).

[6] Collins (1986) and Huneman (2019) argue that this episode was indeed instrumental in introducing the evolutionary viewpoint in ecology.

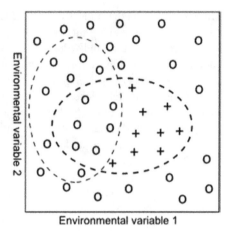

Fig. 13.2 Realised and fundamental niches (the circle dots species is a better competitor than the crosses species) (Hutchinson (1957))

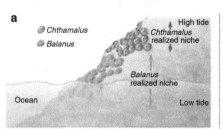
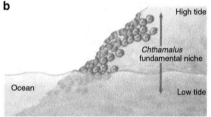

Fig. 13.3 (**a**) Coexistence of Balanus and Chtamallus. (**b**) When Balanus are taken out, Chtamalus reveal their fundamental niche by occupying it

niche. Partitioning the total environmental hyperspace into realised niches through the competitive exclusion principle eventually explains coexistence. The basic idea is that similar species cannot coexist for a long period of time, so one of them ultimately restricts the realised niche of the other: this idea yields a family of "limiting similarity" theories of coexistence that have been elaborated since the 60s. "The generalization (..) that two species with the same niche requirements cannot form mixed steady-state populations in the same region has become one of the chief foundations of modern ecology." (Hutchinson 1957)

As Hutchinson noticed, this theoretical tool is however not able to wholly explain coexistence. The "paradox of plankton", as he himself formulated (Hutchinson 1961), is the fact that while the hyperspace of environmental parameters in the ocean is very small, since there are few parameters (light, pH, temperature, etc.) distributed over a small range of values, thousands of plankton species exist instead of a few ones as predicts the theory of the niche. Hutchinson considered various accounts to explain this, especially the view that in reality the parameter values vary at the same time as the competitive exclusion process operates, which entails that the equilibrium partition of the niche hyperspace predicted by the principle of competitive exclusion cannot be reached,[7] and many more species actually exist.

The niche theory elaborated by Hutchinson has been crucial to address the coexistence issue. Hutchinson (1957) used it to explain why there are so many animals: the huge variety of plants, on which animals feed, makes for a very large hyperspace for possible animals, and hence many realised niches and species. But another crucial ecological aspect is the distribution of species abundances, which is not taken into account by Hutchinson's theory. As seen by Fisher and then Preston in the 30s, the species abundance distribution (SAD) seems to realize constant patterns, which can be expressed by either a lognormal curve (Fisher et al. 1943) or a logseries (Preston 1948). Witnessing such a regularity in SADs, across various kinds of ecosystems on the planet, raised the question of explaining such patterns. Consequently, theories of coexistence approached by limiting similarity elaborated since the 60s have been refined to understand such patterns of biodiversity as SADs. A most recent elaboration of this theory is called the R* theory. Stemming from MacArthur and Levins (1967) paper on patchy environments and the competition between species foraging a finite set of identical resources heterogeneously distributed, and later developed by Tilman (1982), this theory asserts that "when resources are heterogeneously distributed, the number of species can be larger than the number of limiting resources, thereby resolving Hutchinson's paradox of the plankton. R* theory is a conceptual advance over previous phenomenological-competition theories, such as the Lotka–Volterra predator–prey model, because it predicts the outcome of competition experiments before they are performed." (Marquet et al. 2014).

Biodiversity as an *explanandum* means more than the coexistence question and SAD patterns (see e.g. MacArthur 1972). In fact, another pattern discovered by ecologists carrying out censuses (especially on plants) was about the relation of surface area and species number. Called "species-area law", this was also a major pattern to be explained. In the seminal book by MacArthur and Wilson, *Theory of Island Biogeography* (1967), the authors recognise that the distribution of species is a major *explanandum*. However, their aim consists in switching from the "natural history" of species, which is mostly collecting patterns of those distributions, to a mechanistic explanation (more on this below). *Theory of Island Biogeography* starts by elaborating the "inland-island model". Islands are small territories separated from one another by the sea, and all of them are at some distance from the inland. The inland constitutes a reservoir for species. Individuals of those species colonise islands, but the chances of colonizing an island depend both on the distance of inland to X and on the size of X. The mathematical model therefore intends to

---

[7] "At any point the illuminated zone of the ocean or a lake the phytoplankton is normally quite diversified. There is no opportunity for niche specialization and the fundamental trophic requirements of all forms will cause them to draw on the same food supply. Such population cannot therefore represent equilibria, but since in general the plankton, though continually changing, remains in a highly diversified state, one can only suppose that the direction of competition to continually undergoing change with the progress of the seasons and concomitant thermal and chemical changes in the water and that no opportunity for the establishment of a single species equilibrium condition ever occurs." (Hutchinson 1959)

explain varieties of species-area laws on the basis of these three parameters: island size, amount of islands, distances to inland.

Theory of Island Biogeography uses a hypothetico-deductive model-based method: like Fisher or Kimura's population genetics, it starts by building models and then considers how data and patterns fit to the model. *Theory of Island Biogeography* also works at a higher scale than community ecology. "Islands" of course are theoretical entities, not physical islands; they correspond to territories that are poorly communicating genes and organisms to other territories: they can be valleys, forest patches, etc. (Island biogeography has theoretical affinities with the concept of "metapopulation" elaborated at the same time by Richard Levins (1969)).

Noticeably, most of *Theory of Island Biogeography* considers the dispersion and colonisation of species and not the relative fitness differences between individuals. Hubbell (2001, 2006, 2010) will consider this as the first "dispersal assembly" model of coexistence, and will contrast it with the "niche assembly" models deriving from the limiting similarity theory (Leigh 2007). His own theory, called the "neutral unified theory of biogeography and ecology" (Hubbell 2001), intends to elaborate a dispersal assembly model, which is therefore 'neutral' in the sense that, like Kimura's evolutionary models (Kimura 1985; He and Hubbell 2005), there is no fitness differences between elements (alleles for Kimura, species for Hubbell). His model integrates both regional and local community scales, and therefore allows to explain species abundance distribution as well as species-area laws. The key change for his theory (as compared to *Theory of Island Biogeography*) is that the neutrality assumption, called "ecological equivalence", is defined in terms of per capita birth and death rates rather than in terms of species fitness (as McArthur and Wilson were doing) (Munoz and Huneman 2016, Hubbell 2005; Purves and Turnbull 2010). It met predictive success: "It was surprising to find that spatial neutral models give rise to frequency distributions of precision that are very similar to those estimated from biological surveys, as a consequence of the spatial patterns produced by local dispersal alone" (Bell et al. 2006) – which concerned several kinds of communities (see McGill et al. 2006): plants, coral reefs or fish (Muneepeerakul et al. 2008).

Thus, a major divide in contemporary community ecology is today defined by the meaning ascribed to the neutral theory: whether ecologists follow Hubbell in considering that it is a good theory for biodiversity (Bell 2000), especially because it has far less parameters than the rival R* theory (Marquet et al. 2014), or think that the niche paradigm is still the best explanation since neutral models are not explanations (see amongst others Chave 2004; Hubbell 2005; Holt 2006; Allouche and Kadmon 2009; Leibold and McPeek 2006; Doncaster 2009; Rosindell et al. 2011). Without delving into the controversy, it has to be noted that the neutral theory appears as a unified, scale-encompassing theory of biodiversity while the limiting similarity paradigm proposes explanations that are generally different for several aspects of coexistence (species abundance distribution, species area laws, etc.) (Huneman 2017).

In all those theories, the *explanandum* is a range of patterns of coexistence that are defined mostly in terms of *species diversity*, and often the species richness is the major aspect – even though species abundances are also taken crucially into account in SADs.

But what happens if we turn to diversity as an *explanans*?

13.5 Diversity as an *Explanans*

In this section, I will consider the so-called diversity-stability hypothesis and the particular notions of diversity that have been involved in the attempts to clarify, formulate and test this hypothesis for four decades. I will first consider approaches that focus on the way diversity as species richness is organized through interactions in ecological networks; then I shall turn to the notions labelled "functional diversity" and "phylogenetic diversity".

The diversity-stability hypothesis has been for a long time an assumed but unproven hypothesis, evidenced by many observations – somewhat like famous mathematical conjectures that are not proven but seem established by the behaviours of known numbers. As Orians writes, "The belief that natural ecosystems become more diverse and, hence, more stable with time after a disturbance is widely accepted and regularly repeated in ecology textbooks (…) the correlations, not to mention causations, are still obscure." (Orians 1975, p. 139) Indeed, Orians notices that even the correlation between diversity and stability could not exclude that such a common cause as environmental constancy yields both.[8]

To this extent, the real meaning of the terms involved (which diversity? what stability?) was not really investigated.[9] Thus, as indicated above, May's mathematical findings that diversity *per se*, crudely defined as the number of species interacting in an ecosystem, does not beget stability, triggered a reflexive turn in the study of diversity as a stability-promoter.[10]

Robert May's results were in fact showing that an ecosystem with randomly interacting species is less stable – in the sense of constancy of species' abundances – while the amount of species increases. Yet some evidence of a stability-begetting effect of diversity existed in the field, as stated, for example by McNaughton (1977): "The weight of evidence resulting from explicit tests of the diversity-stability hypothesis (…) suggests, not that the hypothesis is invalid, but that it is correct".

---

[8] "Environmental constancy facilitates diversity while reducing perturbations that might affect stability" (Orians 1975, p. 139)

[9] "The concepts are normally discussed with poorly defined terms, reflecting an uncertainty about what concept(s) of stability are useful in ecology" (ib.)

[10] Notice that the regulation issue, arguably another key issue of theoretical ecology, also concerns an aspect of stability, since it is about the steadiness of one species population abundance. Therefore, the diversity as *explanans* is involved in the other major question of ecology, provided one assumes that the key questions are regulation and coexistence.

Thus, ecologists then enquired about what aspects of diversity were in those cases accounting for this effect. First, in contrast with May's models, the connections in an ecological network are often not random. It may be that many predators have only one prey, for instance, and that a few superpredators have many preys. In any case, few ecosystems are such that species have an even chance of having a given number of preys or predators. Thus, the question switched towards the identification of the properties of ecological networks that would be such that increasing diversity would increase stability. Given a fixed degree of diversity as species richness, many networks are possible: a first rough characterization of their differences is their particular value of connectance, namely, the ratio of the amount of realised connections (here, interactions) between species to the total amount of possible connections.

More generally, a perspective on the question of the role of ecological diversity in stability is the general investigation of topological properties of graphs realised by ecological networks of interactions. Diversity, as species richness *per se*, does not increase stability but some network topologies make it likely to promote stability: this hypothesis supports the general move towards an investigation of ecological networks and their role in stability (Solé et al. 2002; Dunne 2006; Dunne et al. 2002; Kéfi et al. 2016). Some of the results emphasize the key role of species networks topologies in guaranteeing some stability. Scale-free networks, in which the distribution of the degrees[11] of the nodes follows some power law,[12] are stable because this topology entails a very low probability for a random species extinction to reach one of the hubs of the network and hence alter the overall structure, and ultimately the functioning of the community (Solé et al. 2002). This probability becomes lower with the increasing size of the network, i.e. with the increase in species richness.

Small-world structures[13] of ecological networks, when they are realised, also beget stability. This is because the high clustering coefficient means that the overall pattern of interaction is mostly preserved if some cluster in the network is altered. On the other hand, the short path length means that a species which loses its privileged interacting species in its neighbouring cluster can still be related to its other interacting species via the other species in its network, to which it is highly connected (Strogatz 2001; Solé and Goodwin 1988). In this case, similarly, increasing

---

[11] In a given network, made up of nodes (or vertices) and edges that connect some nodes, the "degree" of a node is the amount of edges on this node.

[12] Intuitively, there are a few nodes with many connections (they will be called hubs), slightly more hubs with a bit less connections, and so on, and a large majority of nodes with only very few connections. Formally speaking, the number of nodes of degree $n + 1$ will be $1/10$ the number of nodes of degree n. (Or any mathematically power law of the same kind). Wealth in human societies is known to follow power laws; and one frequent generating process for power law nodes distributions is the "preferential attachment", namely, the probability of having a new connection is proportional to the extant amount of connections. Sometimes called "rich get richer", this process is clearly instantiated by financial mechanisms (Albert and Barabasi 2002).

[13] Small-world is a kind of network characterized by the fact that it is highly clustered (a cluster being a set of nodes more significantly connected between themselves than to other nodes) and at the same time has a short path length (the path length being the average number of edges between two randomly taken nodes) (Watts and Strogatz 1998).

the amount of species, hence the size of the network, strengthens this stability-enhancing property.

May's counterintuitive findings about stability not yielded by diversity in general are therefore corrected or supplemented by those network analyses of the topology of ecological network; however, it is not clear exactly what is meant in both cases by "stability". Thus, the meaning of "stability" in all these models had to be questioned. As Tilman (1994) made clear, even if species richness does not, in theory, beget stability as constancy of species *abundances*, it has a positive effect on the constancy of *biomass* of an ecosystem. That is clearly another meaning of "stability", which relies on diversity. And early on, Holling (1973) had introduced "resilience" understood as the ability of an ecological system to restore its key parameters after a perturbation. Resilience has various modes and can be empirically measured. Moreover, "persistence" named the fact that an ecosystem does not "lose" a species, even though the abundances of all species vary a lot and do not come back to the initial state.

Yet, notions of stability are themselves even more numerous than that and it is not even clear if there is one overarching meaning. Orians (1975) distinguishes: *Constancy* – "a lack of change in some parameter of a system, such as the number of species, taxonomic composition, life form structure of a community, or feature of the physical environment"; *Persistence* – "the survival time of a system or some component of it"; *Inertia* – "the ability of a system to resist external perturbations"; *Elasticity* – "the speed with which the system returns to its former state following a perturbation" (which is similar to Holling's resilience); *Amplitude* – "the area over which a system is stable"; *Cyclical Stability* – "the property of a system to cycle or oscillate around some central point or zone"; *Trajectory Stability* – "the property of a system to move towards some final end point or zone despite differences in starting points". (Fig. 13.4) Stability, in other words, depends on the kind of perturbations one considers, and for Orians, in addition, all measures should be related to fitness: "For these relationships to be insightful, perturbations [or perturbation types] should be related to the evolutionary histories of the organisms experiencing the perturbations, and measured in terms of the total investments that must be made to increase or maintain fitness during those perturbations." (ib. p. 143) This indicates a bias in favor of evolutionary approaches to ecology, which may not be found in other theories of stability, especially when one turns to functional or ecosystems ecology.

In any case, the question of which diversity begets stability, and how it is possible that a certain diversity begets a certain stability, presupposes that one clarifies which stability is at stake. Not all diversity properties are likely to beget the same stability property.[14]

Of course, the ecological networks can be understood also from the perspective of their dynamics (Ulanowicz 1983; Szyrmer and Ulanowicz 1987), and especially

---

[14] On the various meanings of stability in ecology, and the possibility of formally making sense of some of them in the context of phase spaces, attractors and measures of Lyapounov exponents, see Justus (2008).

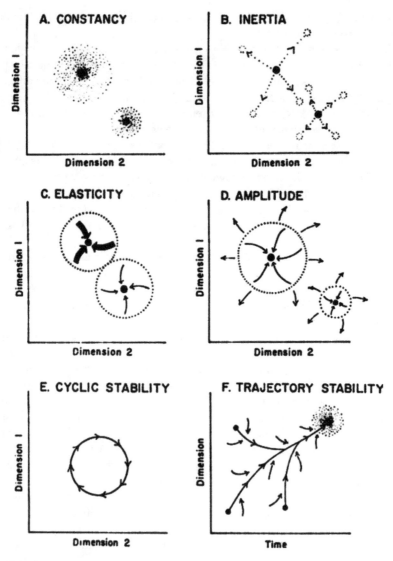

Fig. 13.4 The six kinds of stability in ecology. (After Orians 1975)

by considering not only the constraints put by the topology on the possible dynamics (Huneman 2015), but also, by capturing the major behaviours of the dynamics of fluxes within the networks and the possible evolution of the networks likely to follow (Ulanowicz 1986). This approach is perfectly compatible with a functional ecology that would consider ecosystems as open thermodynamic systems and model their inner behaviour, such as what Odum (1953) theorised. It allows researchers to understand the role that increasing diversity (as species richness) plays in the productivity of ecosystems, or ecosystem functioning, or some key features of ecosystem functioning.

The network perspective is not the only way to capture the possible contributions of diversity, mostly as species richness, to stability, or to some aspects of it. Functional ecologists started to define "functional differences" understood in terms of functional roles of a species played in an ecosystem (Blandin 2014). From this perspective, two species can be biologically different but functionally equivalent. Such functional diversity may be likely to play a role that species diversity cannot play in the emergence and maintenance of some stability. However, functional diversity and species richness are not wholly orthogonal. As Tilman (1996) argued, species diversity induces a lot of microscale environmental heterogeneities, which in turn allow for a wide variety of ecological roles. But this connection is just plausible and does not allow one to always consider species richness as a proxy for functional diversity.

Experiments have recently confirmed the stability-enhancing role of functional diversity. The bumphead parrotfish *Bolbometopon muricatum* is the largest parrotfish in the oceans and is considered a keystone species in the coral reef (Huey and Belwood 2009). It is a major target for fishermen (and hence an imperilled species) but is also heavily consuming reef substrate: "the most conspicuous and perhaps most powerful effect *B. muricatum* has on reef ecology is delivered via individuals' intense direct consumption of reef substrate." (McCauley et al. 2014)

The experimental change of this parrotfish to another parrotfish, or the reintroduction of other parrotfish species after its removal, show that the equilibrium of the coral reef is threatened. Species diversity in this case is not changed (Bellwood et al. 2003). But given that the functional role of the bumphead parrotfish is unique, it follows that functional diversity is decreased while species richness remains constant. In this case functional diversity, and not species diversity, is what contributes to ecosystem stability.

Functional diversity seems thus to positively relate to productivity and stability of ecosystems. However, as argued by Cadotte et al. (2009) "functional group richness is a problematic measure for two reasons. First, the removal or addition of "functionally redundant" species may have effects on community dynamics and processes, indicating that there are important functional differences not captured by broad groupings. (...) The second reason is that functional group richness tends to predict only a limited amount of variation in productivity and may even explain less variation than having randomly assigned groups."

Thus, more recently ecologists have started to consider *phylogenetic* diversity and its role in ecosystem functioning and conservation biology, under the name of "ecophylogenetics". Here, phylogenetic diversity is understood as "the amount of evolutionary history represented in the species of a particular community", and "commonly used measures of phylogenetic diversity are the total branch length of a phylogenetic tree that contains all species present in a community, or the sum of pairwise distances between species weighed by their relative abundances." (Mouquet et al. 2012) Ecologists found, for instance, that plant productivity is enhanced in communities with phylogenetically distantly related fungal species compared to closely related species. "This result suggests, under the hypothesis of a strong phylogenetic signal of the traits considered, that the loss of an entire lineage could have

strong negative ecological consequences since distinct lineages are likely to perform different functions." Thus, to this extent one can use phylogenetic diversity "as
a proxy of unmeasured functional diversity for the purpose of assessing its connection to ecosystem functioning" (Mouquet et al. 2012).

The three diversities, species richness, phylogenetic diversity and functional
diversity, are in general quite decoupled. This is manifest in a study by D'Agata
et al. (2014) on the human impact on biodiversity loss in coral reefs. In the reef area,
human density varies on a gradient spanning from 1,7 to 1720 inhabitants/km².
The researchers investigated the effect of this density upon the three biodiversities.
It turned out that the impact starts to be sensible at a threshold of around 20 inhabit-
ants/km²; however, the effect is very different regarding each kind of diversity.
Considering the extreme impact, at 1705 inhabitants/km² the effects are: on species
richness: 12%; on functional diversity: 46%; on phylogenetic diversity: 36%. Thus,
first, species richness is a very bad predictor of human impact on biodiversity loss
and should be not used as an indicator for coral management, one should prefer
functional and phylogenetic biodiversity instead; second, the slope of the impact
after the threshold, on each diversity, is significantly different, therefore they cannot
be taken as proxies for each other (Fig. 13.5).

Fig. 13.5 Differential
effects on human density
in three kinds of
biodiversity. (After
D'Agata et al. (2014))

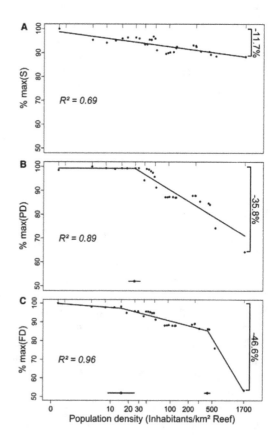

To sum up, diversity as an *explanans* is diffracted into several concepts of diversity such that each plays, within different explanatory perspectives, a specific explanatory role regarding productivity, stability and other ecosystem functioning aspects. Those diversities are not translatable and are in general weakly correlated, even though locally under some conditions they can be quite aligned.

It seems therefore that the explanatory reversibility of diversity includes a gap between the *explanandum* and the *explanans*, since the *explanandum* is mostly concentrated upon species richness, unlike the *explanans*. In turn, the *explanandum* is instantiated in various *patterns* of biodiversity that may link space and diversity, while the *explanans* generally does not include biodiversity patterns (or at least the same biodiversity patterns: SADs, species-area distributions etc.). Two general conclusions can be drawn here: as an *explanans*, ecological diversity is much more diffracted than as an *explanandum*; and the explanatory reversibility of the concept is not transparent, complete or univocal.

One can usefully compare this explanatory reversibility to the explanatory reversibility of robustness in evolutionary biology. Here, robustness, understood either as a capacity to function notwithstanding disturbances, or as an ability to maintain a set of functions in a very wide range of circumstances (Kitano 2004) also covers distinct meanings. Especially, the two key types of robustness for evolutionary biologists are "mutational robustness", as a robustness defined with regard to genetic mutations, and "environmental robustness", as a robustness defined with regard to environmental changes (de Visser et al. 2003). Biologists debate about whether one has been the effect of the other, and then, given that robustness is a very general property of living systems at all levels (Wagner 2005), they ask two kinds of questions: what made robustness evolve (robustness as an *explanandum*)? What does robustness do in evolution and how does it affect it (robustness as an *explanans*)? But such explanatory reversibility of robustness (Huneman 2018) is such that the two types of robustness are together considered, both, in the *explanans* side and in the *explanandum* side. This is not the case with the biodiversity concept in ecology. In the last section, I shall attempt to account for the structure of the concept of diversity in a way that will make sense of this specific explanatory reversibility of the concept. Ultimately, this will decide upon the role of "biodiversity" as a crucial concept for unifying ecology.

13.6   A "Conceptual Space" Approach to the Diversity Concept

What do I mean when I say that some X – a community or an ecosystem – is more diverse than Y? Does it include more species, or species more diverse, or more functionally diversified, or is X phylogenetically more extended on the tree of life than Y?

No principled way exists to answer this question. One could be tempted to say that there is no objective answer at all. However, another approach consists in saying that there are many objective facts enveloped in a judgment about X being more diverse than Y, and that the concept of biodiversity is then in each case built or constructed upon this set of objective facts. Various answers to the question are then yielded by various ways of constructing this concept of biodiversity.

Such an approach could be developed in the following terms: consider each of the properties used to construct biodiversity indices and to measure biodiversity as axes in a hyperspace. Species richness would obviously be one, as would then be species evenness, disparity, species abundance, phylogenetic distance, functional differences. Those axes describe facts about each community or ecosystem that can be objectively measured: the number of species at the local scale, their abundances, the functional redundancies or the amount of the phylogenetic trees covered by the species in a community or metacommunity are not in the eye of the beholder, they can be settled independently of epistemic preferences, explanatory strategies or methodological choices (or, at least, their objectivity is not different or less objective than generally establishing facts in science). Thus each community or ecosystem occupies a point (or a small neighbourhood, considering that the values evolve in time) in this space, defined by how much it scores on each of these axes (Fig. 13.6). Functional diversity is the projection of this point on the axis "functional diversity"; same for phylogenetic diversity; etc.

But of course each axis may not be as important as the others regarding a given diversity measure – for instance, some concepts of diversity used in conservation biology would overtone functional diversity or species abundances; and diversity in ecophylogenetics, but also in biogeography, could overemphasise the axis of "phylogenetic diversity". Many diversity indices are indeed constructed by considering the values on

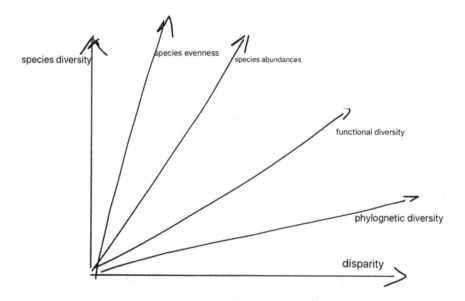

Fig. 13.6 The conceptual space of biodiversity and its axes. Notice that axes can be weighed and scaled differently in order to construct a specific biodiversity concept

several axes but not all of them, and then by possibly weighing the various axes differently; e.g. species richness or disparity could be differently weighed. If one wants to represent in our conceptual space the way a specific biodiversity concept, and then biodiversity measure, is constructed, one could assign different scales to each axis. This plurality of choices regarding the importance, weight or scales of each axis results in a plurality of possible concepts of biodiversity. And in turn, each explanatory project in ecology regarding diversity – as an *explanans*, or as an *explanandum* – will involve one (or a few) specific biodiversity concepts among this plurality.

In this approach, "biodiversity" appears as a possible construction built upon the objective values that X (community, ecosystem) scores on various axes. Each way of constructing it, by making projections on some axes, or taking only a few axes, possibly scaled or weighed in different ways, provides a different concept of biodiversity. Each of these concepts in turn is based on objective facts, but includes some epistemic and possibly *non-epistemic* values that governed the construction of this concept from those facts. For instance, the biodiversity concept used in conservation may emphasize the dimension of abundance, since the probability of extinction of a species – which is in general something conservation biologists intend to prevent – is inversely proportional to abundance. But the weighing of the axes here, and the overweighing of abundances, relies on the non-epistemic value of our interest in conserving species. Inversely, some biodiversity concepts used when one wants to design, maintain or maximise ecosystem services, may favor the functional diversity; here too, the reasons for weighing axes differently relies on non-epistemic values, namely our interest in flourishing ecosystem services.

Now, the explanatory reversibility of the concept of diversity can be approached in this context. Considering that biodiversity is defined in this conceptual space determined by the axes I mentioned, it appears that diversity as *explanans* and diversity as *explanandum* target different regions of this space (Fig. 13.7). According to analyses in Sects. 13.4 and 13.5, the *explanans* is heavily concentrated around the axes on functional and phylogenetic diversity, while the *explanandum* would be rather located around the axes of species richness, evenness and abundances. The overall conceptual space of diversity is therefore not identically involved in the two explanatory takes on diversity, and this characterizes the epistemic nature of such an explanatory reversibility, as compared to the explanatory reversibility of the concept of robustness mentioned above. The latter is "complete", while the former is not – in the sense that the conceptual space (respectively, of diversity and of robustness) is in the latter case completely and identically concerned by both explanatory projects, and in the former, partially and differently concerned by each explanatory project. But (unlike diversity) robustness cannot claim to be a shared and pervasive object in evolutionary biology, and therefore the "completeness" of its explanatory reversibility does not carry consequences for the question of the theoretical unity of evolutionary biology, unlike in the case of ecological diversity considered here.

This approach to diversity as a conceptual space was not only intended to provide a representation for the incompleteness of the explanatory reversibility of diversity. It is more generally intended to make sense of the fact that the epistemic status of diversity in ecology does not allow for a theoretical unity based on such

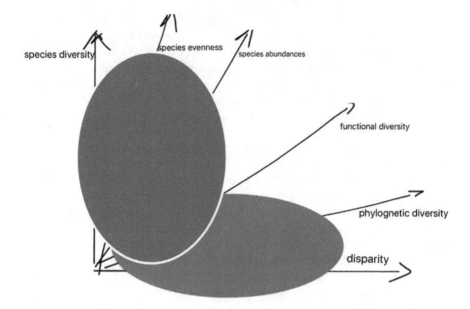

Fig. 13.7 Regions of the conceptual space of biodiversity targeted by explanatory projects: in red, when biodiversity is the *explanans*, in blue, when it is the *explanandum* (Color figure online)

concept, because various explanatory questions in ecology target different conceptual areas of this space. Thus, even though there is here some unity, due to the fact that there is one conceptual space, it is the unity conferred by a same general object. But the fact that the space is so to say differently exploited by various approaches and traditions makes it difficult to think that for this sole reason a theoretical unity can embrace all those approaches and traditions.

Through this "conceptual space approach" to diversity, one sees that diversity is not a purely subjective property, or a property that only exists is the eye of the (scientific) beholder; that many crucial explanatory projects in ecology diversely target diversity; and that at the same time, all these projects cannot be theoretically unified through this reference to diversity as something objective.

13.7   Conclusion

To wrap up the arguments made here, biodiversity is arguably a key issue in ecology, and many theories and explanatory strategies are concerned by it. Diversity is manifestly an explanatory reversible concept, and at first stake this could mean that it could play a role in unifying ecological theories and tools. However, because of the specific incompleteness of the explanatory reversibility of the concept of diversity, illustrated by the way in which its conceptual space is variously targeted by explanatory projects, it appears that the discourses, theories and explanatory strategies of ecologists could not be theoretically unified as a set of scientific approaches to diversity. Even if it is perhaps illusory to think that a single concept could unify a theoretical field, in a non-superficial manner at least, the present enquiry shows that in order to search for unifying principles for ecology, one should not start by focusing on biodiversity. Ecologists share a concern for biodiversity, but the geography of the concept of diversity is such that this shared concern cannot become a principle of unification or an essential part of a unifying strategy.

Acknowledgements I am grateful to the audience of the Urbino conference on biodiversity, organized by Elena Casetta and Andrea Borghini in 2015, where a first version of this chapter has been presented. I warmly thank Davide Vecchi and Elena Casetta, whose comments significantly improved the manuscript, as well as Matt Chew, Sébastien Dutreuil, Antoine Dussault and two anonymous reviewers. This work was possible thanks to the CNRS Laboratoire International Associé ECIEB Paris-Montréal and the GDR "Les savoirs de l'environnemental" (CNRS GDR3770 Sapienv).

References

Albert, R., & Barabasi, A.-L. (2002). Statistical mechanics of complex networks. *Reviews of Modern Physics, 74*, 47–97.

Allee, W. C., Park, O., Emerson, A. E., Park, T., & Schmidt, K. P. (1949). *Principles of animal ecology*. Philadelphia: W. B. Saunders Company.

Allouche, O., & Kadmon, R. (2009). A general framework for neutral models of community dynamics. *Ecology Letters, 12*, 1287–1297.

Andrewartha, H. G., & Birch, L. C. (1954). *The distribution and abundance of animals*. Chicago: University of Chicago Press.

Antonovics, J. (1976). Plant population biology at the crossroads. Input from population genetics. *Systematic Botany, l*, 234–245.

Bell, G. (2000). The distribution of abundance in neutral communities. *The American Naturalist, 155*, 606–617.

Bell, G., Lechowicz, M. J., & Waterway, M. J. (2006). The comparative evidence relating to the neutral theory of community ecology. *Ecology, 87*, 1378–1386.

Bellwood, D. R., Hoey, A. S., & Choat, J. H. (2003). Limited functional redundancy in high diversity systems: Resilience and ecosystem function on coral reefs. *Ecology Letters, 6*, 281–285.

Blandin, P. (2014). La notion de biodiversité: sémantique et épistémologie. In E. Casetta & J. Delord (Eds.), *La biodiversité en questions* (pp. 31–82). Paris: Matériologiques.

Cadotte, M. W., Cavender-Bares, J., Tilman, D., & Oakley, T. H. (2009). Using phylogenetic, functional and trait diversity to understand patterns of plant community productivity. *PLoS One, 4*, e5695.

CBD. (1992). *Convention on biological diversity*, United Nations. http://www.cbd.int/doc/legal/cbd-en.pdf.

Chave, J. (2004). Neutral theory and community ecology. *Ecology Letters, 7*, 241–253.

Chew, M. K. (2011). Invasion Biology: Historical Precedents. In D. Simberloff & M. Rejmánek (Eds.), *Encyclopedia of biological invasions* (pp. 369–375). Berkeley: University of California Press.

Clements, F. E. (1916). *Plant succession* (Carnegie Institute Pubi. no. 242). Washington, DC: Carnegie Institute.

Clements, F., & Shelford, V. (1939). *Bio-ecology*. New York: J. Wiley & Sons, Inc.; London.

Collins, J. P. (1986). Evolutionary ecology and the use of natural selection in ecological theory. *Journal of the History of Biology, 19*, 257–288.

Connell, J. (1961). The influence of interspecific competition and other factors on the distribution of the barnacle chthamalus stellatus. *Ecology, 42*(4), 710–723.

D'Agata, S., Mouillot ,D., Kulbicki, M., Andrefouet, S, Bellwood, D., Cinner, J., Cowman, P., Kronen, M., Pinca, S., Vigliola, L. (2014). Human-mediated loss of phylogenetic and functional diversity in coral reef fishes. *Current Biology, 24*(5), 555–560.

de Visser, J. A. G. M., Hermisson, J., Wagner, G. P., Ancel Meyers, L., Bagheri-Chaichian, H., Blanchard, J. L., & Chao, L. (2003). Evolution and detection of genetic robustness. *Evolution, 57*, 1959–1972.

Doncaster, C. P. (2009). Ecological equivalence: A realistic assumption for niche theory as a testable alternative to neutral theory. *PLoS One, 4*, e7460.

Dunne, J. (2006). The network structure of food webs. In M. Pascual & J. Dunne (Eds.), *Ecological networks: Linking structure to dynamics in food webs*. Oxford: Oxford University Press.

Dunne, J. E., Williams, R. J., & Martinez, N. D. (2002). Food web structure and network theory: The role of connectance and size. *PNAS, 99*, 12917–12922.

Dussault, A., & Bouchard, F. (2017). A persistence enhancing propensity account of ecological function to explain ecosystem evolution. *Synthese, 194*, 1115.

Elton, C. (1927). *The ecology of animals*. New York: Wiley.

Fisher, R. (1932). The evolutionary modification of genetic phenomena. *Proceedings of the 6th International Congress of Genetics, 1*, 165–172.

Fisher, R. A., Corbet, A. S., & Williams, C. B. (1943). The relation between the number of species and the number of individuals in a random sample from an animal population. *Journal of Animal Ecology, 12*, 42–58.

Ford E.B. (1964). *Ecological genetics*. London : Chapman and Hall.

Gleason, H. A. (1939). The individualistic concept of the plant community. *The American Midland Naturalist, 21*, 92–110.

Gosselin, F. (2014). Diversité du vivant et crise d'extinction: des ambiguïtés persistantes. In E. Casetta & J. Delord (Eds.), *La biodiversité en questions* (pp. 119–138). Paris: Matériologiques.

Hagen, J. (1989). Research perspectives and the anomalous status of Modern Ecology. *Biology and Philosophy, 4*, 433–455.

Hagen, J. (1992). *The Entangled bank. The origins of ecosystem ecology*. New Brunswick: Rutgers University Press.

Harper, J. L. (1967). A Darwinian approach to plant ecology. *Journal of Ecology, 55*, 247–270.

Holling, G. (1973). Resilience and stability of ecological systems. *Annual Review of Ecology and Systematics, 4*, 1–23.

Holt, R. D. (2006). Emergent neutrality. *Trends in Ecology & Evolution, 21*(10), 531–533.

Horn, H. S. (1975). The ecology of secondary succession. *Annual Review of Ecology and Systematics, 5*, 25–37.

Hubbell, S. P. (2001). *The unified neutral theory of biodiversity and biogeography*. Princeton: Princeton University Press.

Hubbell, S. P. (2005). Neutral theory in community ecology and the hypothesis of functional equivalence. *Functional Ecology, 19*, 166–172.

Hubbell, S. P. (2006). Neutral theory and the evolution of ecological equivalence. *Ecology, 87*, 1387–1398. 31.

Hubbell, S. P. (2010). Neutral theory and the theory of island biogeography. In J. B. Losos & R. E. Ricklefs (Eds.), *The theory of island biogeography revisited* (pp. 264–292). Princeton: Princeton University Press.

Huey, J., & Bellwood, D. (2009). Limited functional redundancy in a high diversity system: Single species dominates key ecological process on coral reefs. *Ecosystems, 12*, 1316–1328.

Huneman, P. (2015). Diversifying the picture of explanations in biological sciences: Ways of combining topology with mechanisms. *Synthese, 195*, 115–146.

Huneman, P. (2019). "How the Modern Synthesis came to ecology." *Journal of the History of Biology*. Forthcoming, 52, 4.

Huneman, P. (2017). Stephen Hubbell and the paramount power of randomness in ecology. In
O. Harman & M. Dietrich (Eds.), *Dreamers, visionaries and revolutionaries in the life sciences*. Chicago: University of Chicago Press, pp. 176–195.

Huneman, P. (2018). Robustness as an explanandum and explanans in evolutionary biology and
ecology. In M. Bertolaso, S. Caianiello & E. Serelli (Eds.), *Biological Robustness. Emerging
Perspectives from within the Life Sciences*. Dordrecht: Springer, pp. 95–121.

Huneman, P., & Walsh, D. (Eds.). (2017). *Challenging the modern synthesis: Adaptation, development and inheritance*. New York: Oxford University Press.

Hutchinson, G. E. (1957). Concluding remarks. *Cold Spring Harbor Symposia on Quantitative
Biology, 22*, 415–427.

Hutchinson, G. E. (1959). Homage to Santa Rosalia or why are there so many kinds of animals.
American Naturalist, 93, 145–159.

Hutchinson, G. E. (1961). The paradox of the plankton. *American Naturalist, 95*, 137–145.

Ives, R., & Carpenter, J. (2007). Stability and diversity of ecosystems. *Science, 317*(5834), 58–62.

Justus, J. (2008). Ecological and Lyapunov stability. *Philosophy of Science, 75*, 421.

Kéfi, S., Miele, V., Wieters, E. A., Navarrete, S. A., & Berlow, E. L. (2016). How structured is the
Entangled bank? The surprisingly simple organization of multiplex ecological networks leads
to increased persistence and resilience. *PLoS Biology, 14*(8), e1002527.

Kimura, M. (1985). *The neutral theory of molecular evolution*. Cambridge: Cambridge University
Press.

Kingsland, S. (1995). *Modeling nature: Episodes in the history of population ecology* (2nd ed.).
Chicago: University of Chicago Press.

Kitano, H. (2004). Biological robustness. *Nature Reviews Genetics, 5*, 826–837.

Lack, D. (1947). The significance of clutch size. *Ibis, 89*, 302–352.

Lack, D. (1954). *The natural regulation of animal numbers*. Oxford: Oxford University Press.

Leibold, M. A., & McPeek, M. A. (2006). Coexistence of the niche and neutral perspectives in
community ecology. *Ecology, 87*, 1399–1410.

Leigh, E. G. (2007). Neutral theory: A historical perspective. *Journal of Evolutionary Biology, 20*,
2075–2091.

Levins, R. (1969). Some Demographic and Genetic Consequences of Environmental Heterogeneity
for Biological Control. *Bulletin of the Entomological Society of America, 15*(3), 237–240.

Loreau, M. (2010). Linking biodiversity and ecosystems: Towards a unifying ecological theory.
Philosophical Transactions of the Royal Society B, 365(1537), 49–60.

Lortie, C. J., Brooker, R. W., Choler, P., Kikvidze, Z., Michalet, R., Pugnaire, F. I., & Callaway,
R. M. (2004). Rethinking plant community theory. *Oikos, 107*, 433–438.

MacArthur, R. H. (1972). Coexistence of species. In J. A. Behnke (Ed.), *Challenging biological
problems* (pp. 253–259). New York: Oxford University Press.

MacArthur, R., & Levins, R. (1967). The limiting similarity, convergence, and divergence of coexisting species. *American Naturalist, 101*, 377–385.

MacArthur, R. E., & Wilson, E. O. (1967). *The Theory of island biogeography*. Princeton:
Princeton University Press.

Margalef, R. (1963). On certain unifying principles in ecology. *American Naturalist, 97*, 357–374.

Marquet, P., Allen, A., Brown, J., Dunne, J., Enquist, B., Gilloly, J., Gowaty, P. A., Green, J., Harte,
J., Hubbell, S. P., O'Dwyer, J., Okie, J., Ostling, A., Ritchie, M., Storch, D., & West, G. (2014).
On theory in ecology. *Bioscience, 64*, 701–710.

May, R. M. (1974). *Stability and complexity in model ecosystems*. Princeton: Princeton University
Press.

May, R. (1975). Stability in ecosystems: Some comments. In W. H. van Dobben & R. H. Lowe-
McConnell (Eds.), *Unifying concepts in ecology. Report of the plenary sessions of the first
international congress of ecology, The Hague, the Netherlands, 1974* (pp. 161–168). Dordrecht:
Springer.

Mccauley, D. J., Young, H. S., Guevara, R., Williams, G. J., Power, E. A., Dunbar, R. B., Bird,
D. W., Durham, W. H., & Micheli, F. (2014). Positive and negative effects of a threatened parrotfish on reef ecosystems. *Conservation Biology, 28*, 1312–1321.

McGill, B., Maurer, B. A., & Weiser, M. D. (2006). Empirical evaluation of neutral theory. *Ecology, 87*, 1411–1423.

McIntosh, R. P. (1986). *The background of ecology. Concept and theory*. Cambridge: Cambridge University Press.

McNaughton, S. J. (1977). Diversity and stability of ecological communities: A comment on the role of empiricism in ecology. *American Naturalist, 111*(979), 515–525.

Mitman, G. (1988). From the population to society: The cooperative metaphors of W.C. Allee and A.E. Emerson. *Journal of the History of Biology, 21*, 173–192.

Mouquet, N., Devictor, V., Meynard, C. N., Munoz, F., Bersier, L.-F., Chave, J., Couteron, P., Dalecky, A., Fontaine, C., Gravel, D., Hardy, O. J., Jabot, F., Lavergne, S., Leibold, M., Mouillot, D., Münkemüller, T., Pavoine, S., Prinzing, A., Rodrigues, A. S. L., Rohr, R. P., Thébault, E., & Thuiller, W. (2012). Ecophylogenetics: Advances and perspectives. *Biological Reviews, 87*, 769–785.

Muneepeerakul, R., Bertuzzo, E., Lynch, H. J., Fagan, W. F., Rinaldo, A., & Rodriguez-Iturbe, I. (2008). Neutral metacommunity models predict fish diversity patterns in Mississippi-Missouri basin. *Nature, 453*, 220–222.

Munoz, F., & Huneman, P. (2016). From the neutral theory to a comprehensive and multiscale theory of ecological equivalence. *Quarterly Review of Biology, 91*(3), 321–342.

Nicholson, A. J. (1933). 'The Balance of Animal Populations'. *Journal of Animal Ecology, 2*, 132–178.

Nicholson, A., & Bailey, V. (1935). "The Balance of Animal Populations-Part 1". *Proceedings of the Zoological Society*: 551–598.

Nicoglou, A. (2015). The evolution of phenotypic plasticity: Genealogy of a debate in genetics. *Studies in History and Philosophy of Biological and Biomedical Sciences C, 50*, 67.

Noss, R. F. (1990). Indicators for monitoring biodiversity: A hierarchical approach. *Conservation Biology, 4*(4), 355–364.

Odum, E. P. (1953). *Fundamentals of ecology*. Philadelphia: W. B. Saunders Co.

Orians, G. (1975). Diversity, stability and maturity in natural ecosystems. In W. H. van Dobben & R. H. Lowe-McConnell (Eds.), *Unifying concepts in ecology. Report of the plenary sessions of the first international congress of ecology, The Hague, the Netherlands, 1974* (pp. 139–150). Dordrecht: Springer.

Pimm, S. L. (1984). The complexity and stability of ecosystems. *Nature, 307*, 321–326.

Pimm, S. L. (1993). *The balance of nature? Ecological issues in the conservation of species and communities*. Chicago: University of Chicago Press.

Pocheville, A. (2015). The ecological niche: History & recent controversies. In T. Heams, P. Huneman, G. Lecointre, & M. Silberstein (Eds.), *Handbook of evolutionary thinking in the sciences*. Dordrecht: Springer.

Preston, F. W. (1948). The commonness and rarity of species. *Ecology, 29*, 254–283.

Purves, D. W., & Turnbull, L. A. (2010). Different but equal: The implausible assumption at the heart of neutral theory. *Journal of Animal Ecology, 79*, 1215–1225.

Richardson, D. (Ed.). (2011). *Fifty years of invasion ecology: The legacy of Charles Elton*. London: Wiley.

Rosindell, J., Hubbell, S. P., & Etienne, R. S. (2011). The unified neutral theory of biodiversity and biogeography at age ten. *Trends in Ecology and Evolution, 26*, 340–348.

Roughgarden, J. (1979). *Theory of population genetics and evolutionary ecology: An introduction*. London: Prentice Hall.

Schoener, T. W. (1972). Mathematical ecology and its place among the sciences. *Science, 178*, 389–391.

Servedio, M. R., Brandvain, Y., Dhole, S., Fitzpatrick, C. L., Goldberg, E. E., et al. (2014). Not just a theory–The utility of mathematical models in evolutionary biology. *PLoS Biology, 12*(12), e1002017.

Slack, N. (2010). *G. Evelyn Hutchinson and the invention of modern ecology*. New Haven: Yale University Press.

Smith, F. E. (1952). Experimental methods in population dynamics. *Ecology, 33*, 441–450.

Solé, R., & Goodwin, B. (1988). *Signs of life: How complexity pervades biology*. New York: Basic Book.

Solé, R. V., Ferrer, R., Montoya, J. M., & Valverde, S. (2002). Selection, tinkering and emergence in complex networks. *Complexity, 8*, 20–33.

Solomon, M. E. (1949). The natural control of animal population. *Journal of Animal Ecology, 18*, 1: 1–1:34.

Stearns, S. (1982). The emergence of evolutionary and community ecology as experimental science. *Perspectives in Biology and Medicine, 25*(4), 621–648.

Strogatz, S. (2001). Exploring complex networks. *Nature, 410*, 268–276.

Szyrmer, J., & Ulanowicz, R. E. (1987). Total flows in ecosystems. *Ecological Modelling, 35*, 123–136.

Takacs, D. (1996). *The Idea of Biodiversity. Philosophies of Paradise*. Baltimore and London, John Hopkins University Press.

Tilman, D. (1982). *Resource competition and community structure*. Princeton: Princeton University Press.

Tilman, D. (1994). Competition and biodiversity in spatially structured habitats. *Ecology, 75*, 2–16.

Tilman, D. (1996). Biodiversity: Population versus ecosystem stability. *Ecology, 77*, 350–363.

Ulanowicz, R. E. (1983). Identifying the structure of cycling in ecosystems. *Mathematical Biosciences, 65*, 219–237.

Ulanowicz, R. E. (1986). *Growth and development: Ecosystems phenomenology*. New York: Springer.

Vellend, M. (2010). Conceptual synthesis in community ecology. *Quarterly Review of Biology, 85*, 183–206.

Vellend, M. (2016). *The theory of ecological communities*. Princeton: Princeton University Press.

Volterra, V. (1926). Variazioni e fluttuarzioni del numero d'individui in specie animali conviventi. *Memoria della Reale Accademia Nazionale dei Lincei, 2*, 31–113.

Wagner, A. (2005). *Robustness and evolvability in living systems*. Princeton: Princeton University Press.

Warming, E. (1909). *Oecology of plants*. Oxford: Clarendon Press.

Watts, D. J., & Strogatz, S. H. (1998). Collective dynamics in "small-world" networks. *Nature, 393*, 440.

Williams, G. C. (1966). *Adaptation and natural selection*. Princeton: Princeton University Press.

Wilson, E. O. (1992). *The diversity of life*. Cambridge, MA: Belknap Press

Wilson, E. O. (Ed.). (1988). *Biodiversity*. Washington, DC: National Academy Press.

14

The Concept of Functional Diversity

Antoine C. Dussault

Abstract This chapter argues that the common claim that the ascription of ecological functions to organisms in functional ecology raises issues about levels of natural selection is ill-founded. This claim, I maintain, mistakenly assumes that the function concept as understood in functional ecology aligns with the selected effect theory of function advocated by many philosophers of biology (sometimes called "The Standard Line" on functions). After exploring the implications of Wilson and Sober's defence of multilevel selection for the prospects of defending a selected effect account of ecological functions, I identify three main ways in which functional ecology's understanding of the function concept diverges from the selected effect theory. Specifically, I argue (1) that functional ecology conceives ecological functions as *context-based* rather than *history-based* properties of organisms; (2) that it attributes to the ecological function concept the aim of explaining ecosystem processes rather than that of explaining the presence of organisms within ecosystems; and (3) that it conceives the ecological functions of organisms as *use* and *service* functions rather than *design* functions. I then discuss the extent to which the recently proposed causal role and organizational accounts of ecological functions better accord with the purposes for which the function concept is used in functional ecology.

Keywords Functional biodiversity · Function · Selected effect theory · Ecosystem selection · Superorganism

A. C. Dussault (✉)
Centre interuniversitaire de recherche sur la science et la technologie (CIRST), Université du Québec à Montréal (UQAM), Montréal, QC, Canada

Département de philosophie, Collège Lionel-Groulx, Sainte-Thérèse, QC, Canada
e-mail: antoine.cdussault@clg.qc.ca

14.1   Introduction

In the last decades, *functional biodiversity* has become a central focus in ecology and environmental conservation (e.g. Tilman 2001; Naeem 2002; Petchey and Gaston 2006; Nock et al. 2016). This follows from the recognition by an increasing number of ecologists of the explanatory and predictive limitations of more traditional "species richness" measures of biodiversity. This recognition has led ecologists and conservationists to consider, alongside the number of species present in a community, the particular features of organisms of those species and how those features determine their potential relationships with their environments (see Hooper et al. 2002, 195; DeLaplante and Picasso 2011, 173; Nunes-Neto et al. 2016, 296–297). Consideration of those features has fostered among ecologists an interest in the ways in which organisms can be grouped or classified on the basis of their *functional traits*, which are deemed to be of more direct ecological importance than those on which the more standard taxonomic measures of biodiversity are based.

Those functional groupings include:

Guilds: Groupings of organisms on the basis of similarities in resource use. Two organisms are members of a same guild if they tend to use a similar resource in a similar way (Simberloff and Dayan 1991; J. B. Wilson 1999; Blondel 2003).

Functional response groups: Groupings of organisms on the basis of similar expected response to environmental changes. Two organisms are members of the same functional response group if they tend to respond similarly to similar changes in environmental conditions (Catovsky 1998; J. B. Wilson 1999; Hooper et al. 2002; Lavorel and Garnier 2002)

Functional effect groups: Groupings of organisms on the basis of similar roles in ecosystem processes. Two organisms are members of the same functional effect group if they tend to contribute similarly to some important ecosystem process (e.g. nutrient cycling, primary productivity, energy flows) (Catovsky 1998; Hooper et al. 2002; Lavorel and Garnier 2002; Blondel 2003).[1]

Among those three modes of functional classification, the first two—*guilds* and *functional response groups*—are commonly used to explain the assembly of ecological communities and how their species composition changes in response to changes in their environments. The third—*functional effect groups*—is commonly used to explain

---

[1] It should be noted that functional ecologists have adopted various modes of functional classification with different emphases, and have used diverse terminologies to refer to them. For instance Wilson (1999) draws a contrast between *alpha guilds* and *beta guilds* which is essentially equivalent to the contrast made above between *guilds* and *functional response groups*. Similarly, Catovsky (1998), and Lavorel and Garnier (2002) draw a contrast between *functional response groups* and *functional effect groups* similar to the one made above, but define *functional response groups* also in reference to resource use (a basis for classification that I associated with *guilds*). And likewise, Blondel (2003) draws a contrast between *guilds* and *functional groups*, and his concept of *functional group* is essentially equivalent to the above concept of *functional effect group*. I think that my above identification of three main modes of functional classification adequately reflects the complementary epistemic aims in relation to which ecologists use functional classifications.

ecosystem processes through delineating the particular contributions of organisms of different species to those processes (see discussion in Sect. 14.3.2 below).

A particularity of the third mode of functional classification—*functional effect groups*—is that it involves the ascription of *roles* or *functions* to organisms within ecosystems (Catovsky 1998, 126; Symstad 2002, 23–24; Jax 2010, 54). As remarked by Jax (2010, sec. 4.2) and DeLaplante and Picasso (2011, sec. 3.2), such ascriptions of ecological functions to organisms within ecosystems raise important philosophical issues. One of them concerns the meaning of the *function* concept and its relationship to claims about natural selection. Given the association made by many biologists and ecologists between the concept of function and the evolutionary concept of *adaptation* (Williams 1966), the idea that organisms fulfil functions within ecosystems has been claimed to raise issues about the levels at which natural selection customarily operates (see Calow 1987, 60; DeLaplante and Picasso 2011, 184). As we shall see, a linkage of the notion of ecological function to community and ecosystem selection assumes an elucidation of this notion along the lines of the *selection effect* theory of function advocated by many philosophers of biology (e.g. Wright 1973; Millikan 1989; Neander 1991; Godfrey-Smith 1994).[2] According to this theory, which some refer to as "The Standard Line" on functions given its many adherents (Allen and Bekoff 1995, 13–14), the function of a part or trait of a biological entity is the effect for which this part or trait was preserved by natural selection operating on the ancestors of that entity. A selected effect elucidation of the concept of ecological function would therefore entail that ascribing a function to an organism within an ecosystem amounts to saying that at least some of the traits of this organism have been shaped by ecosystem-level selection. Relatedly, a selected effect elucidation of the ecological function concept, as we shall also see, would in some way revive the old idea of communities and ecosystems as tightly integrated *superorganisms* shaped by natural selection (Allee et al. 1949; D. S. Wilson and Sober 1989).

In this chapter, I will argue that the common association between function ascriptions in functional ecology and issues about levels of selection is ill-founded. As just mentioned, this association assumes an understanding of ecological functions along the lines of the *selection effect* theory of function, and I will maintain that the understanding of the function concept at play in functional ecology does not in fact align with this theory. I will do so through identifying important ways in which functional ecology's use of the ecological function concept diverges from the understanding conveyed by the selected effect theory. This will highlight that, when they ascribe functions to organisms within ecosystems, functional ecologists are not committed to views of ecosystems as units of selection. Their understanding of ecological functions and ecosystem functional organization, as I will emphasize, attributes to ecosystems a lower degree of part-whole integration than what would be entailed by the selected effect theory. The discussion of the ecological function concept presented in this chapter will therefore reinforce the near consensus that has recently emerged among philosophers of biology and ecology, according to which the ecological

---

[2] For overviews of philosophical theories of function, see McLaughlin (2001), Wouters (2005), Walsh (2008), Saborido (2014), and Garson (2016).

function concept should be elucidated along the lines of non-selectionist alternatives to the selected effect theory of function (Maclaurin and Sterelny 2008, sec. 6.2; Odenbaugh 2010; Gayon 2013; Nunes-Neto et al. 2014).

My discussion will be organized as follows. In Sect. 14.2, I will discuss the common contention that the use of the function concept in ecology raises issues about levels of selection. I will explore the implications of Wilson and Sober's defence of multilevel selection for the prospects of defending a selected effect account of ecological functions. In Sect. 14.3, I will dispute the claim that the ecological function concept raises issues about levels of natural selection. I will do so by highlighting three important ways in which functional ecology's understanding of the ecological function concept diverges from the selected effect theory. Finally, in Sect. 14.4, I will briefly discuss two non-selectionist accounts of ecological functions that have recently been proposed by philosophers of biology and ecology; namely, the *causal role* account (Maclaurin and Sterelny 2008, sec. 6.2; Odenbaugh 2010; Gayon 2013), and the *organizational* account (Nunes-Neto et al. 2014). I will maintain that neither of these two accounts fully accords with how ecological functions are understood in functional ecology.

14.2   Ecological Functions and Levels of Selection

As mentioned in the introduction, the ascription of ecological functions to organisms in functional ecology is often taken to raise issues about levels of natural selection. As DeLaplante and Picasso (2011, 184) recall:

> [A]ttitudes toward function language in ecology have been influenced by the group selection debate that took place in the 1960s (Wynne-Edwards 1962; Williams 1966). The critique of group selection was based on the affirmation that within orthodox evolutionary theory, natural selection acts primarily at the level of individual organisms (or, indeed, the level of individual genes), and rarely if ever at the level of groups. [...] Evolutionary ecologists tend to associate the language of functions with organism-environment relationships relevant to selection and adaptation (*e.g.*, "functional traits"). But if natural selection only acts at the level of individuals within species populations, then the language of functions should only apply at this level [...]. Consequently, evolutionary ecologists are inclined to be skeptical of function attributions at the community and ecosystem level.[3]

Along similar lines, in the inaugural issue of the journal *Functional Ecology*, Calow (1987, 60) maintains that a focus on the functions fulfilled by organisms within communities "implies that the way they contribute to the balanced economy of the community is an important criterion of selection".

Such a linkage of the notion of ecological function to community or ecosystem selection assumes an understanding of this notion along the lines of the *selected effect* theory of function developed in the philosophy of biology (Wright 1973;

---

[3] For are more detailed discussion of the issues raised by the group selection debate for functional approaches to ecology, see Hagen (1992, chap. 8).

Millikan 1989; Neander 1991). Some support for this assumption can be found in the fact that the selected effect theory has, to some extent, established itself as "The Standard Line" on functions in the philosophy of biology (Allen and Bekoff 1995, 13–14). Since its initial introduction, it has been adopted by many prominent philosophers of biology (e.g. Griffiths 1993; Mitchell 1993; Godfrey-Smith 1994). According to the selected effect theory, the function of a part or trait of a biological entity is the effect for which this part or trait was preserved by natural selection operating on ancestors of this entity. Thus, ascribing a function to an organism within an ecosystem would amount to saying that at least some of the traits of this organism have been shaped by ecosystem-level selection. In other words, ascribing a function to an organism within an ecosystem would amount to saying that organisms from its lineage have the traits on account of which they are classified in a particular *functional effect group* partly because their having those traits conferred a selective advantage to the ecosystem they are part of. Thus, functional ecologists' ascribing ecological functions to organisms within ecosystems would commit them to the idea that communities and ecosystems are units of natural selection. The view of ecosystem functional organization implicitly adopted in functional ecology would therefore be similar to that espoused by mid-Twentieth century ecologists who believed that communities and ecosystems were tightly integrated *superorganisms* subject to community or ecosystem-level selection (e.g. Allee et al. 1949).

Although, as remarked by DeLaplante and Picasso (see quote above), many biologists and ecologists are sceptical about the idea that natural selection customarily operates at the level of communities and ecosystems, some support for this idea can be found (as they also remark) in Wilson and Sober's defence of multilevel selection (see e.g. Wilson and Sober 1989; Sober and Wilson 1994). Wilson and Sober's main focus is population-level selection, but they also apply their multilevel selectionist approach to communities and ecosystems. Wilson and Sober's defence of multilevel selection improves upon previous defences in part by identifying an unrealistic assumption underlying classical arguments against it. This assumption is that individual organisms within populations interact randomly with each other and therefore have equal chances of mating with any other member of their population. Contrary to this assumption, Wilson and Sober emphasize, the heterogeneity of many environments entails that, in practice, populations in the ecological world tend to be structured in ways that make their individual members more likely to interact with only a small subset of their whole population. This, as Wilson and Sober explain, creates conditions favourable to the operation of natural selection on single-species groups of organisms and even communities and ecosystems (Wilson and Sober 1989, 341–4).

They illustrate the possibility of community-level selection with the example of *phoretic associations*. Phoretic associations are communities formed by a winged insect associated with many wingless organisms (e.g. mites, nematodes, fungi and microbes) that rely on the winged insect for transportation from one resource patch to another. When the winged insect reaches a new resource patch (e.g. carrion, dung, or stressed timber), it brings along a whole community of "phoretic associates" which then colonize the patch. Wilson and Sober explain how natural selection might operate on phoretic associations as a whole:

> Consider a large number of resource patches, each of which develops into a community composed of the insects, their phoretic associates, plus other species that arrive independently. The community of phoretic associates may be expected to vary from patch to patch in species composition and in the genetic composition of the component species. Some of these variant communities may have the effect of killing the carrier insect. Others may have the effect of promoting insect survival and reproduction, and these will be differentially dispersed to future resource patches. Thus, between-community selection favors phoretic communities that do not harm and perhaps even benefit the insect carrier. At the extreme, we might expect the community to become organized into an elaborate mutualistic network that protects the insect from its natural enemies, gathers its food, and so on. (Wilson and Sober 1989, pp. 348–9)

Such a scenario, they emphasize, is not only a theoretical possibility. Empirical data from studied phoretic communities show no negative effects on the carrier insect in most cases and positive effects in many cases. In a subsequent paper, Wilson (1997, 2020–22) discusses other likely cases of community selection that conform to his and Sober's approach, as well as a likely case of ecosystem selection involving micro-ecosystems forming at the surface of lakes and oceans.[4]

Wilson and Sober's defence of community and ecosystem selection thus seems to provide grounds for interpreting at least some of the functions fulfilled by organisms within communities and ecosystems along the lines of the selected effect theory of function. For instance, the selected effect theory entails that some phoretic associates in Wilson and Sober's phoretic association case have functions within the phoretic association. This is the case of phoretic associates that are part of the association partly because some of their traits conferred a selective advantage to the phoretic association as a whole. Similar function ascriptions would be implied by the selected effect theory in relation to organisms involved in the other cases of community and ecosystem selection described by Wilson (1997). In line with those observations, Wilson and Sober themselves conceive their defence of multilevel selection as legitimizing the view that some communities and ecosystems are *functionally organized* entities (Wilson and Sober 1989, 337–344; see also Wilson 1997). They even claim that communities and ecosystems that are units of selection according to their approach can genuinely be regarded as *superorganisms* (Wilson and Sober 1989, 349).[5]

However, it should be emphasized that Wilson and Sober's defence of multilevel selection lends at best very limited support to the application of the selected effect theory in ecology. Wilson and Sober are careful to emphasize that their defence of community and ecosystem selection is professedly modest. They see it as an important strength of their approach that it does not consist in an "overly grandiose" superorganism theory that attributes "functional design [...] to ecosystems in general"

---

[4] For related discussions of artificial ecosystem selection experiments, see Swenson et al. (2000a), Swenson et al. (2000b) and Blouin et al. (2015).

[5] For a discussion of Wilson and Sober's defence of multilevel selection in relation to the selected effect theory of function, see Basl (2017, sec. 4.2).

(Wilson and Sober 1989, 352). As they insist, their approach entails that "[n]ot all groups and communities are superorganisms, but only those that meet the specified (and often stringent) conditions" (Wilson and Sober 1989, 343). Functional ecologists, in contrast, envision their approach as a framework for the study of ecosystems in general. Such a broad scope is not legitimized by Wilson and Sober's approach. Therefore the support lent by Wilson and Sober's defence of multilevel selection to the application of the selected effect theory of function in ecology seems too limited for the purposes of functional ecology.

In the next section, I will argue that significant aspects of the use of the function concept in functional ecology point to an understanding of function that diverges from the selected effect theory. This will show that, contrary to what is sometimes suggested (see above), the ascription of ecological functions to organisms in functional ecology does not hinge on claims that ecosystems are units of natural selection.

14.3   Ecological Functions in Functional Ecology

14.3.1   Ecological Context vs. Selective History

Historically and conceptually, contemporary functional ecology's construal of the function concept derives from the renowned community ecologist Charles Elton's (1927, 1933) understanding of the *ecological niche*. Elton's understanding of the niche was tied to a functionalist view of ecological communities, which drew an analogy between feeding interactions within ecological communities and economic exchanges in human societies.[6] In Elton's coinage, the term "niche" referred to "what [an animal] is *doing* in its community", and emphasized an animal's "*relations to food and enemies*" in contrast to "appearance, names, affinities, and past history." (Elton 1927, 63–64, emphasis in the original) The niche concept was "used in ecology in the sense that we speak of trades or *professions* or *jobs* in a human community" (Elton 1933, 28, emphasis added). Thus, Elton's understanding of the niche was tied to a picture of ecological communities in analogy with human societies (with an economic focus), rather than with individual organisms. The niches of organisms, as he conceived them, were analogous to the economic roles fulfilled by individuals within human societies, rather than with the functions of organs within organisms. This communitarian-economic analogy attributed to ecological commu-

---

[6] Elton's understanding of the niche contrasted with the one previously adopted by Joseph Grinnell (1917), the other originator of the niche concept, who used the niche concept to denote a species' particular *environmental requirements* (see Leibold 1995, 1372–1373). The contrast between Grinnell's and Elton's niches parallels the contrast presented in the introduction between on the one hand, *guilds* and *functional response group* and on the other hand, *functional effect groups* (see Hooper et al. 2002, 196). For discussions of the contrast between Grinnell's and Elton's niche concepts, see also Schoener (1989), Griesemer (1992), and Pocheville (2015).

nities of a lower degree of part-whole integration than the one characteristically found in individual organisms. Notably, Elton (1930) emphasized that individual organisms retain a significant degree of autonomy with respect to the communities in which they are involved, and he rejected the view (held by some later Twentieth-century ecologists) that natural selection customarily operates on ecological communities as a whole (McIntosh 1985, 167; Haak 2000, 32).

Contemporary functional ecology's understanding of ecological functions is in many respects similar to Elton's functional understanding of the niche. A first important aspect of this understanding that does not align with the selected effect theory concerns the basis on which ecological functions are ascribed to organisms in functional ecology. In functional ecology, the ecological functions of organisms within ecosystems are conceived as *context-based* properties of those organisms, which they bear on account of their actual and potential interactions with other organisms. This context-based understanding contrasts with that conveyed by the selected effect theory, according to which the functions of biological items are *history-based* properties of those items (i.e. properties borne by those items on account of their selective history). The conceptual dissociation of the ecological function concept from evolutionary considerations is made explicit by some functional ecologists. Petchey and Gaston (2006, 742), for instance, state that "[f]unctional diversity [in ecology] generally involves understanding communities and ecosystems based on what organisms do, rather than on their evolutionary history".

Functional ecology's context-based understanding of ecological functions is aptly portrayed by Jax (2010, 79):

> In contrast to parts of an organism, a particular species has no clearly defined role within an ecosystem: a bird may have the function of being prey to other animals—but only if these carnivorous animals are parts of the specific system. If there are no predators in the system, the same species or even individual will not have the role "prey". Even if we can say that the bird actually has the role of being prey, we can also find other roles, e.g. its role to distribute seeds and nutrients, to be predator for insects, etc. That is, like a person within a human society, who may be teacher, spouse, child, politician etc., either at the same time or at different times, it can have several roles. Roles can change and the same person as well as the same species can even take opposing roles in time […]. "The" one and only role of a species does not exist. Roles are strongly context-dependent.

On this context-based understanding, the ascription of ecological functions to organisms within ecosystems does not entail claims about selective history. For instance, an ecologist's depiction of a rabbit as fulfilling the role of a prey (or primary consumer) within an ecosystem does not entail the claim that rabbits and their traits were selected for serving as food for predators. Rabbits eat grass and grow muscles for their own survival and, as a by-product, acquire traits that make them nutritious and palatable for those predators. Likewise, an ecologist's reference to foxes as fulfilling the role of regulator of herbivore populations within an ecosystem does not entail the claim that foxes and their traits were selected for regulating herbivore populations. Foxes chase and eat preys to feed themselves and, as a by-product, exert a form of control over their preys' populations.

It should be noted, however, that contemporary functional ecology expands upon Elton's approach to the study of ecological communities in two important ways.

First, it expands upon Elton's approach by integrating ecosystem ecology's thermo-dynamic and biogeochemical outlook on the ecological world (see Hagen 1992, chaps. 4–5). Thus, whereas Elton used the niche primarily to study how interspecific interactions within communities explain the regulation of populations within them and the maintenance of their structural features (Hagen 1992, 52; Pocheville 2015, 549), the ascription of ecological functions to organisms in contemporary functional ecology is more primarily tied to the aim of studying how the traits of organisms determine their potential contributions to ecosystem processes (see K. W. Cummins 1974; Naeem 2002). Thus, in contemporary functional ecology, the ecological functions of organisms are their particular contributions to ecosystem processes (e.g. nutrient cycling, primary productivity, energy flows). Contemporary functional ecologists ascribe functions to organisms in order to delineate their particular contribution to the realization and maintenance of those processes.

Second, contemporary functional ecology expands upon Elton's focus on feeding (or trophic) interactions between organisms, by also considering ecological functions acquired by organisms through *non-trophic* interactions with other organisms. Those non-trophic interactions are ones in which organisms affect each other's lives through other means than the direct provision of food (in the form of living or dead tissues). Important non-trophic ecological functions include those fulfilled by *ecosystem engineers*, i.e. organisms that create, modify and maintain habitats in ways that affect the lives of other organisms (e.g. beavers build dams and in so doing create habitats and make many resources available for numerous other organisms) (Jones et al. 1994, 1997; Berke 2010). Non-trophic ecological functions also include those of *pollinators* and *seed dispersers* (see Blondel 2003, 227–228).

Those two significant expansions notwithstanding, it remains the case that ecological function ascriptions as conceived in functional ecology do not involve claims about selective history. For instance, an ecologist's saying that, by building a dam, a beaver fulfils the role of a pond provider with respect to the numerous organisms for which the pond is a favourable habitat does not entail the claim that beavers were selected for providing habitats to those organisms. Beavers build dams and create ponds for their own benefit and, as a by-product, provide habitats to numerous organisms.

An important research aim associated with functional ecology's context-based understanding of function is that of studying the *functional equivalence* between phylogenetically-divergent organisms. Elton (1927, 65), for instance, remarked that the arctic fox, which subsists on guillemot eggs and seal remains left by polar bears, occupies essentially the same niche as the spotted hyæna in tropical Africa, which feeds upon ostrich eggs and zebra remains left by lions. Although they have evolved their traits in distinct selective contexts, arctic foxes and spotted hyæna occupy similar niches. Along similar lines, contemporary functional ecologists have identified functional equivalences, for instance, between ants, birds and rodents, which similarly contribute to seed dispersal in some desert ecosystems, and between hummingbirds, bats and moths, which similarly contribute to the pollination of Lauraceae (a family of plants from the group of angiosperm that usually have the form of trees or shrubs) (see Blondel 2003, 226). The acknowledgement of functional equivalences between phylogenetically-divergent organisms conflicts with the understanding of function conveyed by the selected effect theory, in that this theory would entail that two organisms can have similar ecological functions only to the extent that their traits have evolved in similar selective contexts.

14.3.2 The Explanatory Aim of Ecological Functions

A second important aspect of functional ecology's understanding of functions that diverges from the selected effect theory concerns the *explanatory aim* attributed to the function concept. In functional ecology, as seen in the preceding section, the *explanandum* of ecological function ascriptions is ecosystem processes. The ecological functions of organisms are their particular contributions to the ability of ecosystems to realize and maintain those processes. This contrasts with the *explanadum* of function ascriptions according to the selected effect theory. According to the selected effect theory, the *explanandum* of ecological function ascriptions is the *presence* of the biological items to which functions are ascribed within a system (typically an organism). For instance, according to the selected effect theory, saying that pumping blood is the function of the heart entails not only saying that pumping blood is the way in which hearts contribute to blood circulation in animals with circulatory systems. It also entails saying that animals with circulatory systems have hearts because hearts pump blood (i.e. that hearts *are present* within those organisms because they pump blood). The selected effect functions of biological items explain the presence of those items because, by definition, those functions are the effects for which those items were preserved by natural selection.

To make plain that the *explanadum* of ecological function ascriptions in functional ecology is not the *presence* of organisms within ecosystems, we must recall functional ecology's three main modes of functional classification identified in the introduction. As seen in the introduction, functional ecologists use three main modes of functional classification: (1) *guilds* (groupings based on similar resource use), (2) *functional response groups* (groupings based on similar response to environmental factors), and (3) *functional effect groups* (grouping based on similar roles in ecosystem functioning). As also seen in the introduction, the mode of functional classification that is concerned with functions of organisms within ecosystems is the third one (i.e. *functional effect groups*). However, the modes of functional classification that are primarily involved in the theoretical frameworks used by functional ecologists to explain the presence of organisms within ecosystems are the two other ones (*guilds* and *functional response groups*). Those functional classifications are the ones primarily involved in theories developed for explaining the *assembly* of ecological communities and how communities respond to changes in environmental conditions (through changes in species composition). According to those theories (see Keddy 1992; Díaz et al. 1999), the ability of some particular organisms to establish and maintain themselves in a given community depends, first, on their ability to tolerate the local environmental conditions, and, second, on their ability to exploit the resources available in this community (which requires them to be able to successfully compete with other organisms also using those resources or to share those resources with them). The former ability depends upon the *functional response group* to which organisms belong, and the latter one depends upon their *guild*. The *functional effect groups* to which organisms belong play no significant role in explaining the assembly of ecological communities and their responses to environmental changes.

To be sure, if some regular coincidence could be found between, on the one hand, *guilds* and *functional response groups*, and on the other hand, *functional effects groups,* then one could argue that an explanatory connection nevertheless exists between the ecological functions of organisms and their presence within ecosystems. Functional ecologists, however, emphasize the frequent non-coincidence of those groupings (see e.g. Lavorel and Garnier 2002; Blondel 2003). For instance, birds can disperse some plants' seeds in three different ways: (1) through catching seeds in their plumage and then accidentally dropping them elsewhere (epizoochory), (2) through swallowing fruits and then regurgitating or defecating them elsewhere (endozoochory), or (3) through caching dry fruit seeds for future use and then "forgetting" them (synzoochory). Birds that disperse some plants' seeds in those three ways all belong to the same *functional effect group*. However, insofar as only the birds that disperse seeds in the two latter ways (endozoochory and synzoochory) use the seeds as resources, those birds and those that disperse seeds in the former way (epizoochory) do not belong to the same *guild* (see Blondel 2003, 227–228). Likewise, some varieties of dung beetles feed upon the non-digestive part of large herbivores' green food. Those dung beetles do so in three different ways: (1) through dwelling inside the dung, (2) through burying pieces of the faeces from 0.5 to 1 meter under the dung, and (3) through making a ball of dung, laying eggs within it and rolling it to a place where they can bury it. All dung beetles use the dung as a resource and therefore belong to the same *guild*. However, insofar as the different ways of using the resource lead to different decomposition processes, the three types of dung beetles do not belong to the same *functional effect group* (see Blondel 2003, 228).

It may be objected that the *functional effect groups* to which organisms belong must at least partly explain their presence within ecosystems, given that organisms depend upon the achievement of ecosystem processes for their own existence, and, for this reason, depend, at least indirectly, upon the reliable fulfilment of their own functional contributions to those processes. By fulfilling their ecological functions, in other words, organisms must indirectly contribute to the realization and maintenance of their own conditions for existence, such that they are indirect causes of their continued presence within the ecosystem (or at least of the continued presence of organisms of their *functional effect group*).

I think, however, that this kind of causal link between the fulfilment of their ecological functions by organisms and their presence within ecosystems can, at best, be very weak. Strictly speaking, what organisms contribute to realizing, by fulfilling their ecological functions, is not the conditions necessary for their own presence within an ecosystem, or even for the presence of organisms from their *functional effect group*. What they contribute to realizing is, more accurately, the conditions necessary for the presence of organisms from the *guild* or *functional response group* to which they belong. Abilities to exploit the conditions organisms contribute to realizing by fulfilling their ecological function are determined by membership in *guilds* and *functional response groups*, not by membership in *functional effect groups*. This is well illustrated by a phenomenon studied by ecologists as the "negative selection effect" (Jiang et al. 2008). The "negative selection effect" occurs when some ecological function stops being fulfilled as a result of the displacement of a species that fulfils this function (i.e. that belongs to a particular *functional effect group*) by another species that does not fulfil it (i.e. that does not belong to the same *functional effect group*). The reason why the latter species displaces the former one is that both species use the same resource (i.e. belong to the same *guild*) and the latter species is better at competing for this resource. Thus, suppose, that a species S fulfils the ecological function F within the ecosystem E, and that, by doing so, S contributes to the realization of environmental condition C and to the availability of resource R within E. S therefore belongs to the *functional effect group f* (which encompasses organisms that are able to fulfil F), and also belongs to the *guild r* and the *functional response group c* (which encompass, respectively, organisms that use resource R and that require environmental conditions C). Now, we can see more clearly that, by contributing to the realization of C and the availability of R, organisms from S only weakly promote their own presence (or the presence of other species from *f*) in E. What organisms from S promote by contributing to the realization of C and the availability of R is, in fact, the presence of any species from *guild r* and *functional response group c*. By doing so, therefore, organisms from S promote their own presence within E only provided that there is no other species S_1 that also belongs to c and r and that is more efficient than S in exploiting R. If such a species comes around, then the fulfilment of their ecological function by organisms from S will instead promote the presence of S_1 within the ecosystem, and consequently S's own displacement by S_1. And if S_1 does not belong to f and S was the only species that fulfiled F within E, then F will stop being fulfiled in E. Likewise, by contributing to the realization of C and the availability of R, organisms from S may promote the presence of other species from *functional effect group f* only to the extent that those other species belonging to *f* also belong to r and c. There, however, is no reason to assume that, on a general basis, species that belong to *f* will also belong to r and c. The possibility of such a "negative selection effect," I think, makes clear that the *functional effect groups* to which organisms belong have only limited relevance to the aim of explaining why they are present within ecosystems.

14.3.3 By-Products and the Notion of "Functioning as"

As indicated in Sect. 14.3.1, in functional ecology, ecological functions may be ascribed to organisms on the basis of traits that are evolutionary by-products rather than selected effects (on this point, see also Maclaurin and Sterelny 2008, 115; and

Odenbaugh 2010, 251). This observation points to a third important aspect of functional ecology's understanding of functions that does not align with the selected effect theory. This aspect can be highlighted by drawing the connection between functional ecology's understanding of the function concept and Achinstein's (1977, 350–6) delineation of three distinct meanings of "function" in ordinary language: *design*, *use* and *service* functions. An entity's *design* function consists in what this entity was *designed* or *created* to do (e.g. the function of a mouse trap is to catch mice); whereas an entity's *use* function consists in what it is *used for* (e.g. this table is used for sitting), and an entity's *service* function consists in what it *serves as* (e.g. a watch's second hand serves as a dust sweeper). A table's functioning as a seat or the second hand of a watch's functioning as a dust sweeper do not entail that tables and second hands have been (intentionally) designed for those functions. This distinction between design functions on the one hand, and use and service functions on the other hand, is sometimes also expressed in terms of a contrast between the notion of *being the function of* (e.g. breathing is the function of the nose) and that of *functioning as* (e.g. the nose functions as an eyeglass support) (e.g. Boorse 1976, 76; Bedau 1992, 787–789).

In light of this distinction, the selected effect theory of functions can be interpreted as concerned with *design* functions, that is, as concerned with specifying *the function of* some biological item (as is reflected in selected effect theorists' typical association of function with *design*, see e.g. Wright 1973, 164–65; Millikan 1984, 17). In contrast, functional ecology's context-based functions can be conceived as concerned with *use* and *service* functions, that is, as concerned with specifying what an ecological item can *functions as* in relevant ecological contexts. For instance, rabbits that are preyed upon by foxes in an ecosystem *function as* primary consumers within that ecosystem. In turn, foxes that prey upon those rabbits and exert some control on their population *function as* regulators of the rabbit population within that ecosystem. And likewise, beavers that build dams within an ecosystem and by doing so create habitats and make many resources available for numerous other organisms *function as* pond providers within that ecosystem. Similar to the cases of a table's functioning as a seat and the watch's second hand's functioning as a dust sweeper, rabbits' functioning as primary consumers, foxes' functioning as regulators of rabbit populations and beavers' functioning as pond providers within an ecosystem do not entail claims that rabbits, foxes and beavers were (evolutionarily) designed for fulfilling those functions. Functional ecology thus seems to make use of an ordinary notion of function that is conceptually distinct from the one that the selected effect theory is meant to elucidate. It is not concerned with functions that organisms are (evolutionarily) *designed* to fulfil within ecosystems, but, with functions that they (more fortuitously) fulfil as a result of being (context-dependently) involved in use and service interactions with other organisms.

Above, I maintained that functional ecology attributes to ecological communities a lower degree of part-whole integration than the one characteristically found in individual organisms (in line with Elton's analogy between ecological communities and human societies). Interpreting ecological functions as use and service functions provides some illumination of this idea. A notable feature of individual organisms seems to be their characteristic *teleological integration* (see Queller and Strassmann 2009, 3144). The parts of organisms seem, in some biologically relevant sense, to be *designed* for fulfilling their functions within those organisms. In Achinstein's terminology, the parts of organisms have *design* functions. For instance, hearts do not merely fulfil the role of pumping blood within organisms with circulatory systems, they are (evolutionarily) *designed* for doing so.

Insofar as functional ecology conceives the functions fulfilled by organisms within ecosystems as *use* and *service* functions (in contrast to *design* functions), then functional ecology does not attribute to ecosystems the kind of teleological integration commonly attributed to individual organisms. From the theoretical perspective of functional ecology, ecosystems are functionally organized in a much weaker way than paradigm individual organisms. They are functionally organized not in virtue of being superorganisms shaped by ecosystem-level selective processes, but, more weakly, in virtue of being more or less self-maintaining networks of organisms involved in use and service interactions with each-other. Those use and service interactions collectively generate the ecosystem processes in relation to which functional ecologists ascribe functions to organisms. This view of ecosystem functional organization contrasts with that espoused by mid-Twentieth century ecologists who depicted ecosystems as tightly unified *superorganisms* shaped by community or ecosystem-level natural selection.

14.4 What Is an Ecological Function, Then?

In the previous section, I identified three aspects of functional ecology's understanding of ecological functions that do not align with the selected effect theory of function:

1. Functional ecology conceives ecological functions as *context-based* rather than *history-based* properties of organisms
2. Functional ecology attributes to the ecological function concept the aim of explaining ecosystem processes rather than that of explaining the presence of organisms within ecosystems
3. Functional ecology conceives the ecological functions of organisms as *use* and *service* functions rather than *design* functions

Those three aspects, I think, indicate that, contrary to what is often assumed (see Sect. 14.2), the ascription of ecological functions to organisms in functional ecology does not hinge on claims that natural selection customarily operates at the level of ecosystems. Functional ecology's understanding of the function concept diverges from "The Standard Line" on function according to which functions in biology must be understood as naturally selected effects.

Through highlighting the three aspects just mentioned, the above discussion reinforces the near consensus that has recently emerged among philosophers of biology and ecology, according to which the ecological function concept should be elucidated along the lines of non-selectionist alternatives to the selected effect theory of function (see Nunes-Neto et al. 2013).[7] Philosophers who share this consensus have proposed accounts of ecological functions along the lines of Cummins's (1975) *causal role* theory (Maclaurin and Sterelny 2008, 114–115; Odenbaugh 2010, 251–252; Gayon 2013, 76–77), or along those of Mossio et al. (2009) *organizational* theory of function (Nunes-Neto et al. 2014). How do these accounts stand with respect to functional ecology's use of the function concept?

In some significant respects, the *causal role* theory of function accords with functional ecology's use of the function concept as characterized above. The causal role theory ascribes functions to the parts of biological entities in a way that is entirely independent of their selective history. Function ascriptions, in the causal role theory, serve to identify the particular contributions of the parts of a system to the activities or capacities of that system. This use of the function concept concords with functional ecology's understanding of ecological functions as contributions of organisms to ecosystem processes (see Cooper et al. 2016, sec. 4). Moreover, in line with the above linkage of functional ecology's understanding of functions with Achinstein's notions of *use* and *service* functions (see Sect. 14.3.3), the causal role theory does not confer a privileged epistemic status to the notion of *being the function of* over that of *functioning as* (see Cummins 1975, 762; Craver 2001, 55). Thus, the causal role theory seems to better accord with functional ecology's use of the function concept.

However, a significant limitation of the causal role theory in relation to functional ecology, I think, is its ultimate relativization of functions to the epistemic interests of researchers. According to the causal role theory, parts of a system can be ascribed functions in relation to any capacity or activity of this system that researchers are interested in explaining, provided that the relation between this capacity or activity and the individual contributions of the system's parts is complex enough.[8] As many critics of the causal role theory point out, one problem with this liberal take on functions is that it implausibly entails that functions can be ascribed to the parts of a system on account of their contributions to capacities that amount to deteriorations of those systems (e.g. that a function can be ascribed to a tumour on account of its contribution to the capacity of an organism to die from cancer, see Neander 1991, 181). Thus, on a causal role account, ecological functions could, for instance, be ascribed to organisms from an invasive species on account of their contribution to the ecosystem's capacity to collapse (the fragilization of ecosystems and their possible collapse resulting from the establishment of invasive species is indeed something that ecologists are interested in explaining). Such a degree of inclusive-

---

[7] Dissenters from this consensus are Bouchard (2013) and Dussault and Bouchard (2017), who argue that ecological functions should be understood as contributions to ecosystem fitness (conceived as ecosystem resilience). It should nonetheless be noted that Dussault and Bouchard do not advocate a selected effect account of ecological functions, but rather a forward-looking evolutionary account derived from Bigelow and Pargetter's (1987) dispositional theory of function.

[8] For more details on how causal role theorists substantiate this complexity requirement, see Cummins (1975, 764), Davies (2001, chap. 4), and Craver (2001, sec. 3.2).

ness, I think, does not appropriately reflect the fact that functional ecologists tend to ascribe functions to organisms mainly in relation to capacities or activities of eco-systems that contribute to those ecosystems' ability to maintain themselves. Those processes include primary productivity, nutrient cycling, water uptake, storage of resources, etc. (see, enumerations of ecosystem processes in Walker 1992, 20; and Blondel 2003, 226). Thus, the common objection that the causal role theory is overly liberal also seems to apply in the case of ecological functions.

An organizational account of ecological functions would avoid this problem. The organizational theory defines the functions of the parts of a system as their contribu-tion to the ability of the system to maintain its organization (see Mossio et al. 2009). Such a linkage between functions and the self-maintenance of systems excludes function ascriptions in relation to capacities that amount to deteriorations of sys-tems (see Nunes-Neto et al. 2014, 137–138). In this respect, the organizational theory of function seems to restrict function ascriptions in a way that is consistent with the use of the concept in functional ecology.

However, an important limitation of the organizational theory in relation to func-tional ecology, I think, is that it shares with the selected effect theory the idea that function ascriptions in part explain the presence of function bearers within systems. According to the organizational theory, a biological item has a function within a system if, on the one hand, it contributes to the maintenance of the organization of this system, and if, on the other hand, it is in turn maintained by the organization of the system (Mossio et al. 2009, 16–20). Thus, according to the organizational the-ory, the function bearing parts of a system indirectly contribute to (and therefore explain) their own presence within this system through contributing to that system's maintenance. In this regard, the organizational theory is similar to the selected effect theory (though, in contrast to the selected effect theory, the organizational theory does not make it a requirement that natural selection be the process through which the function bearing parts of systems promote their own presence). The organiza-tional theory therefore attributes to function ascriptions an explanatory aim that is foreign to functional ecology's understanding of the concept. As seen in Sect. 14.3.2, ecological functions as understood in functional ecology are not conceived as explanatory of the presence of organisms within ecosystems. The presence of organisms within ecosystems is explained by their belonging to some *guilds* and *functional response groups*, not by their belonging to some *functional effect groups*. Ecological function ascriptions and the grouping of organisms in *functional effect groups* serve to explain the realization and maintenance of ecosystem processes through delineating the particular contribution of organisms to those processes.

Hence, neither the *causal role* nor the *organizational* account of ecological func-tions fully accord with functional ecology's use of the function concept. The observations made in this section, however, suggest that functional ecology requires an account of functions that combines aspects of those two accounts while eschew-ing some of their other aspects. An elaboration of such an account must be deferred to future work.

14.5 Conclusion

In the preceding sections, I criticised the common supposition that the ascription of ecological functions to organisms in functional ecology hinges on claims that natural selection customarily operates at the level of ecosystems. This supposition, I maintained, rests on the incorrect assumption that the function concept as understood in functional ecology aligns with the selected effect theory of function advocated by many philosophers of biology (sometimes deemed "The Standard Line" on functions). After exploring the implications of Wilson and Sober's defence of multilevel selection for the prospects of defending a selected effect account of ecological functions, I identified three main ways in which functional ecology's understanding of the function concept diverges from the selected effect theory. Specifically, I argued (1) that functional ecology conceives ecological functions as *context-based* rather than *history-based* properties of organisms; (2) that it attributes to the ecological function concept the aim of explaining ecosystem processes rather than with that of explaining the presence of organisms within ecosystems; and (3) that it conceives the ecological functions of organisms as *use* and *service* functions rather than *design* functions. I then briefly discussed the recently proposed accounts of ecological functions along the lines of the causal role and organizational theories of function, and concluded that functional ecology requires an account of functions that selectively draws on those two accounts.

Acknowledgements The author would like to thank Léa Derome, Anne-Marie Gagné-Julien, Philippe Huneman and anonymous referees for valuable comments, as well as O'Neal Buchanan for linguistic revision of the manuscript. The work for this chapter was supported by a postdoctoral fellowship from the Social Sciences and Humanities Research Council of Canada (SSHRC, 756-2015-0748) and a research grant from the Fonds de recherche du Québec – Société et culture (FRQSC, 2018-CH-211053).

References

Achinstein, P. (1977). Function statements. *Philosophy of Science, 44*, 341–367.
Allee, W. C., Emerson, A. E., Park, O., Park, T., & Schmidt, K. P. (1949). *Principles of animal ecology*. Philadelphia: Saunders Co.
Allen, C., & Bekoff, M. (1995). Function, natural design, and animal behavior: Philosophical and ethological considerations. In N. S. Thompson (Ed.), *Perspectives in ethology* (Vol. 11, pp. 1–46). New York: Plenum Press.
Basl, J. (2017). A trilemma for teleological individualism. *Synthese, 194*, 1057–1074. https://doi.org/10.1007/s11229-017-1316-0.
Bedau, M. (1992). Where's the good in teleology? *Philosophy and Phenomenological Research, 52*, 781–806.
Berke, S. K. (2010). Functional groups of ecosystem engineers: A proposed classification with comments on current issues. *Integrative and Comparative Biology, 50*, 147–157. https://doi.org/10.1093/icb/icq077.
Bigelow, J., & Pargetter, R. (1987). Functions. *Journal of Philosophy, 84*, 181–196.
Blondel, J. (2003). Guilds or functional groups: Does it matter? *Oikos, 100*, 223–231. https://doi.org/10.1034/j.1600-0706.2003.12152.x.

Blouin, M., Karimi, B., Mathieu, J., & Lerch, T. Z. (2015). Levels and limits in artificial selection of communities. *Ecology Letters, 18*, 1040–1048. https://doi.org/10.1111/ele.12486.

Boorse, C. (1976). Wright on functions. *Philosophical Review, 85*, 70–86.

Bouchard, F. (2013). How ecosystem evolution strengthens the case for functional pluralism. In P. Huneman (Ed.), *Functions: Selection and mechanisms* (pp. 83–95). Springer: Dordrecht.

Calow, P. (1987). Towards a definition of functional ecology. *Functional Ecology, 1*, 57–61. https://doi.org/10.2307/2389358.

Catovsky, S. (1998). Functional groups: Clarifying our use of the term. *Bulletin of the Ecological Society of America, 79*, 126–127. https://doi.org/10.2307/20168223.

Cooper, G. J., El-Hani, C. N., & Nunes-Neto, N. (2016). Three approaches to the teleological and normative aspects of ecological functions. In N. Eldredge, T. Pievani, E. Serrelli, & I. Tëmkin (Eds.), *Evolutionary theory: A hierarchical perspective* (pp. 103–124). Chicago: University of Chicago Press.

Craver, C. F. (2001). Role functions, mechanisms, and hierarchy. *Philosophy of Science, 68*, 53–74. https://doi.org/10.1086/392866.

Cummins, K. W. (1974). Structure and function of stream ecosystems. *BioScience, 24*, 631–641. https://doi.org/10.2307/1296676.

Cummins, R. C. (1975). Functional analysis. *Journal of Philosophy, 72*, 741–764.

Davies, P. S. (2001). *Norms of nature: Naturalism and the nature of functions*. Cambridge, MA/London: The MIT Press.

DeLaplante, K., & Picasso, V. (2011). The biodiversity–ecosystem function debate in ecology. In K. DeLaplante, B. Brown, & K. A. Peacock (Eds.), *Philosophy of ecology* (pp. 219–250). Oxford/Amsterdam/Waltham: Elsevier.

Díaz, S., Cabido, M., & Casanoves, F. (1999). Functional implications of trait–environment linkages in plant communities. In E. Weiher & P. Keddy (Eds.), *Ecological assembly rules* (pp. 338–362). Cambridge: Cambridge University Press.

Dussault, A. C., & Bouchard, F. (2017). A persistence enhancing propensity account of ecological function to explain ecosystem evolution. *Synthese, 194*, 1115–1145.

Elton, C. S. (1927). *Animal ecology*. New York: The Macmillan Company.

Elton, C. S. (1930). *Animal ecology and evolution*. Oxford: Clarendon Press.

Elton, C. S. (1933). *The ecology of animals* (3rd ed.). London: Methuen.

Garson, J. (2016). *A critical overview of biological functions* (SpringerBriefs in philosophy). Cham: Springer.

Gayon, J. (2013). Does oxygen have a function, or where should the regress of functional ascriptions stop in biology? In P. Huneman (Ed.), *Functions: Selection and mechanisms* (pp. 67–79). Dordrecht: Springer.

Godfrey-Smith, P. (1994). A modern history theory of functions. *Noûs, 28*, 344–362.

Griesemer, J. R. (1992). Niche: Historical perspectives. In E. F. Keller & E. Lloyd (Eds.), *Keywords in evolutionary biology*. Cambridge, MA: Harvard University Press.

Griffiths, P. E. (1993). Functional analysis and proper functions. *British Journal for the Philosophy of Science, 44*, 409–422.

Grinnell, J. (1917). The niche-relationships of the California thrasher. *Auk, 34*, 427–433.

Haak, C. (2000). *The concept of equilibrium in population ecology*. Doctoral dissertation, Halifax, Nova Scotia: Dalhousie University.

Hagen, J. B. (1992). *An entangled bank: The origins of ecosystem ecology*. New Brunswick: Rutgers University Press.

Hooper, D. U., Solan, M., Symstad, A., Diaz, S., Gessner, M. O., Buchmann, N., Degrange, V., et al. (2002). Species diversity, functional diversity and ecosystem functioning. In M. Loreau, S. Naeem, & P. Inchausti (Eds.), *Biodiversity and ecosystem functioning: Synthesis and perspectives* (pp. 195–208). Oxford: Oxford University Press.

Jax, K. (2010). *Ecosystem functioning*. Cambridge/New York: Cambridge University Press.

Jiang, L., Pu, Z., & Nemergut, D. R. (2008). On the importance of the negative selection effect for the relationship between biodiversity and ecosystem functioning. *Oikos, 117*, 488–493. https://doi.org/10.1111/j.0030-1299.2008.16401.x.

Jones, C. G., Lawton, J. H., & Shachak, M. (1994). Organisms as ecosystem engineers. *Oikos, 69*, 373–386.
Jones, C. G., Lawton, J. H., & Shachak, M. (1997). Positive and negative effects of organisms as physical ecosystem engineers. *Ecology, 78*, 1946–1957. https://doi.org/10.1890/0012-9658(1997)078[1946:PANEOO]2.0.CO;2.
Keddy, P. A. (1992). Assembly and response rules: Two goals for predictive community ecology. *Journal of Vegetation Science, 3*, 157–164. https://doi.org/10.2307/3235676.
Lavorel, S., & Garnier, E. (2002). Predicting changes in community composition and ecosystem functioning from plant traits: Revisiting the holy grail. *Functional Ecology, 16*, 545–556. https://doi.org/10.1046/j.1365-2435.2002.00664.x.
Leibold, M. A. (1995). The niche concept revisited: Mechanistic models and community context. *Ecology, 76*, 1371–1382. https://doi.org/10.2307/1938141.
Maclaurin, J., & Sterelny, K. (2008). *What is biodiversity?* Chicago: University of Chicago Press.
McIntosh, R. P. (1985). *The background of ecology: Concept and theory*. Cambridge/New York: Cambridge University Press.
McLaughlin, P. 2001. *What functions explain: Functional explanation and self-reproducing systems* (Cambridge studies in philosophy of biology). Cambridge/New York/Melbourne: Cambridge University Press.
Millikan, R. G. (1984). *Language, Thought, and Other Biological Categories: New Foundations for Realism*. Cambridge, MA/London: MIT Press.
Millikan, R. G. (1989). In defense of proper functions. *Philosophy of Science, 56*, 288–302.
Mitchell, S. D. (1993). Dispositions or etiologies? A comment on Bigelow and Pargetter. *Journal of Philosophy, 60*, 249–259.
Mossio, M., Saborido, C., & Moreno, A. (2009). An organizational account of biological functions. *British Journal for the Philosophy of Science, 60*, 813–841.
Naeem, S. (2002). Functional biodiversity. In H. A. Mooney & J. G. Canadell (Eds.), *Encyclopedia of global environmental change* (pp. 20–36). Chichester/Rexdale: Wiley.
Neander, K. (1991). Functions as selected effects: The conceptual analyst's defense. *Philosophy of Science, 58*, 168–184.
Nock, C. A., Vogt, R. J., & Beisner, B. E. (2016). Functional traits. In *Encyclopedia of life sciences* (pp. 1–8). Chichester: Wiley. https://doi.org/10.1002/9780470015902.a0026282.
Nunes-Neto, N., Moreno, A., & El-Hani, C. N. (2013). The implicit consensus about function in philosophy of ecology. In N. Nunes-Neto, C. N. El-Hani, & A. Moreno (Eds.), *The functional discourse in contemporary ecology* (pp. 40–65). Salvador: Doctoral dissertation, Universidade Federal da Bahia.
Nunes-Neto, N., Moreno, A., & El-Hani, C. N. (2014). Function in ecology: An organizational approach. *Biology and Philosophy, 29*, 123–141.
Nunes-Neto, N., Do Carmo, R. S., & El-Hani, C. N. (2016). Biodiversity and ecosystem functioning: An analysis of the functional discourse in contemporary ecology. *Filosofia e História da Biologia, 11*, 289–321.
Odenbaugh, J. (2010). On the very idea of an ecosystem. In A. Hazlett (Ed.), *New waves in Metaphysics* (pp. 240–258). Basingstoke: Palgrave Macmillan.
Petchey, O. L., & Gaston, K. J. (2006). Functional diversity: Back to basics and looking forward. *Ecology Letters, 9*, 741–758. https://doi.org/10.1111/j.1461-0248.2006.00924.x.
Pocheville, A. (2015). The ecological niche: History and recent controversies. In T. Heams, P. Huneman, G. Lecointre, & M. Silberstein (Eds.), *Handbook of evolutionary thinking in the sciences* (pp. 547–586). Dordrecht: Springer. https://doi.org/10.1007/978-94-017-9014-7_26.
Queller, D. C., & Strassmann, J. E. (2009). Beyond society: The evolution of organismality. *Philosophical Transactions of the Royal Society B: Biological Sciences, 364*, 3143–3155. https://doi.org/10.1098/rstb.2009.0095.
Saborido, C. (2014). New directions in the philosophy of biology: A new taxonomy of functions. In M. C. Galavotti, D. Dieks, W. J. Gonzalez, S. Hartmann, T. Uebel, & M. Weber (Eds.), *New directions in the philosophy of science* (The philosophy of science in a European perspective) (pp. 235–251). Dordrecht: Springer. https://doi.org/10.1007/978-3-319-04382-1_16.

Schoener, T. W. (1989). The ecological niche. In J. M. Cherrett & A. D. Bradshaw (Eds.), *In Ecological concepts: The contribution of ecology to an understanding of the natural world* (pp. 79–114). Oxford: Blackwell Scientific Publications.

Simberloff, D., & Dayan, T. (1991). The guild concept and the structure of ecological communities. *Annual Review of Ecology and Systematics, 22*, 115–143. https://doi.org/10.1146/annurev.es.22.110191.000555.

Sober, E., & Wilson, D. S. (1994). A critical review of philosophical work on the units of selection problem. *Philosophy of Science, 61*, 534–555.

Swenson, W., Arendt, J., & Wilson, D. S. (2000a). Artificial selection of microbial ecosystems for 3-chloroaniline biodegradation. *Environmental Microbiology, 2*, 564–571.

Swenson, W., Wilson, D. S., & Elias, R. (2000b). Artificial ecosystem selection. *Proceedings of the National Academy of Sciences, 97*, 9110–9114. https://doi.org/10.1073/pnas.150237597.

Symstad, A. J. (2002). An overview of ecological plant classification systems. In R. S. Ambasht & N. K. Ambasht (Eds.), *Modern trends in applied terrestrial ecology* (pp. 13–50). New York: Springer.

Tilman, David. 2001. Functional diversity. In Encyclopedia of biodiversity, 3:109–120. Amsterdam Elsevier. doi:https://doi.org/10.1016/B0-12-226865-2/00132-2.

Walker, B. H. (1992). Biodiversity and ecological redundancy. *Conservation Biology, 6*, 18–23.

Walsh, D. M. (2008). Function. In P. Stathis & M. C. London (Eds.), *The Routledge companion to philosophy of science*. New York: Routledge.

Williams, G. C. (1966). *Adaptation and natural selection: A critique of some current evolutionary thought*. Princeton: Princeton University Press.

Wilson, D. S. (1997). Biological communities as functionally organized units. *Ecology, 78*, 2018–2024.

Wilson, J. B. (1999). Guilds, functional types and ecological groups. *Oikos, 86*, 507–522.

Wilson, D. S., & Sober, E. (1989). Reviving the superorganism. *Journal of Theoretical Biology, 136*, 337–356.

Wouters, A. (2005). The function debate in philosophy. *Acta Biotheoretica, 53*, 123–151.

Wright, L. (1973). Functions. *Philosophical Review, 82*, 139–168.

Wynne-Edwards, V. C. (1962). *Animal dispersion in relation to social behavior*. Edinburgh: Oliver and Boyd.

15

Ecosystem Services, Evolutionary Theory and Biodiversity

Silvia Di Marco

Abstract Currently, one of the central arguments in favour of biodiversity conservation is that it is essential for the maintenance of ecosystem services, that is, the benefits that people receive from ecosystems. However, the relationship between ecosystem services and biodiversity is contested and needs clarification. The goal of this chapter is to spell out the interaction and reciprocal influences between conservation science, evolutionary biology, and ecology, in order to understand whether a stronger integration of evolutionary and ecological studies might help clarify the interaction between biodiversity and ecosystem functioning as well as influence biodiversity conservation practices. To this end, the eco-evolutionary feedback theory proposed by David Post and Eric Palkovacs is analysed, arguing that it helps operationalise niche construction theory and develop a more sophisticated understanding of the relationship between ecosystem functioning and biodiversity. Finally, it is proposed that by deepening the integration of ecological and evolutionary factors in our understanding of ecosystem functioning, the eco-evolutionary feedback theory is supportive of an "evolutionary-enlightened management" of biodiversity within the ecosystem services approach.

Keywords Ecosystem functions · Evolution · Niche construction · Ecosystem engineering · Conservation biology

15.1 Introduction

Currently, one of the central arguments in favour of biodiversity conservation is that it is essential for the maintenance of ecosystem services, that is, the benefits that people receive from ecosystems (MA 2003, 2005). However, as remarked by Georgina Mace and colleagues, although both biodiversity and ecosystem scientists implicitly acknowledge that biodiversity plays different roles at the different levels

S. Di Marco (✉)
Centro de Filosofia das Ciências da Universidade de Lisboa, Lisbon, Portugal
e-mail: sdmarco@fc.ul.pt

of the ecosystem services hierarchy, their approach to biodiversity conservation remains fundamentally different. Conservation biologists typically struggle to develop an evidence base that supports the protection of biodiversity, in particular charismatic and endangered species, as a good endowed with cultural, scientific and even "intrinsic" value, while ecologists focus on the contribution provided by biodiversity, usually understood as functional diversity, to ecosystem processes and services (Mace et al. 2012). Face to the challenges posed by the ecosystem services approach to biodiversity conservation, this mismatch amongst professionals is a reason of concern. Still, the growing interest amongst ecologists for the feedbacks between organisms and ecosystems promises to shed new light on the interactions between biodiversity, ecosystem processes and ecosystem services, and has the potential to influence biodiversity conservation planning.

In this regard, various authors stress the fact that since the introduction of the concept of ecosystem service in conservation policies, community and ecosystem ecologists have paid more and more attention to biodiversity, especially species and genes diversity, as a driver of ecosystem functioning (Naeem 2002; Loreau 2010). In particular, Michel Loreau has argued that if ecologists are to understand and model the effects of biodiversity on the functioning of ecosystems, they have to develop new theories to connect the dots that link the evolution of species traits at the individual level (evolutionary biology), the dynamics of species interactions (community ecology) and the overall functioning of ecosystems (ecosystem ecology) (Loreau 2010). An endeavor whose difficulties cannot be understated, especially if one takes into account the "explanatory reversibility" of the concept of biodiversity in ecology,[1] and the philosophical issues posed by both the notion of ecosystem *function* and the idea that organisms play a *role* in an ecosystem.[2]

Bracketing these questions, as well as the problems posed by the polysemy of 'biodiversity',[3] the present chapter aims to spell out the interaction and reciprocal influence between conservation science, evolutionary biology, and ecology, in order to understand whether a stronger integration of evolutionary and ecological studies might help clarify the relationship between biodiversity and ecosystem functioning, and influence biodiversity conservation practices within the ecosystem services approach.

To this aim I will first describe the divide between what Mace et al. (2012) have called the "ecosystem services perspective" and the "conservation perspective" within the ecosystem services approach, and present Loreau's view on the possible integration of ecological and evolutionary studies. Subsequently, I will analyse the eco-evolutionary feedback theory by Post and Palkovacs (2009), as an example of such integration. In particular, I will argue that this theory helps operationalise the evolutionary concept of niche construction (Laland et al. 1999; Odling-Smee et al. 2003), and offers theoretical instruments to develop a more sophisticated understanding of the relationship between ecosystem functioning and biodiversity.

---

[1] See Huneman, Chap. 13, in this volume.

[2] See Dussault, Chap. 14, in this volume.

[3] See Toepfer, Chap. 16, and Meinard et al., Chap. 17, in this volume.

Finally,[4] I will argue that by deepening the integration of ecological and evolutionary factors in our understanding of ecosystem functioning, the eco-evolutionary
feedback theory is supportive of an "evolutionary-enlightened management"
(Ashley et al. 2003) of biodiversity within the ecosystem services approach.

15.2   On the Relationship Between Biodiversity
and Ecosystem Services

Ecosystem services are the benefits that humans derive, directly or indirectly, from
the ecosystems or, phrased differently, they are "the functions and processes of ecosystems that benefit humans" (Costanza et al. 2017). They are classified into *provi-
sioning services*, such as food, clear water, timber, and fuel; *regulating services*,
such as flood protection, pests control, and climate regulation; *supporting services*,
corresponding to basic ecosystem processes such as primary production, soil formation, and nutrients cycle; and *cultural services*, corresponding to a range of cultural
benefits – e.g., aesthetic, recreational, or spiritual – that people receive from
ecosystems.

15.2.1   *Ecosystem Services in Brief*

The idea of ecosystem service is a socio-economic concept that dates back to 1977,
when *Science* published the article "How much are Nature's services worth?" by
Walter Westman, but gained momentum in the academia only in 1997, with the
publication of the book *Nature's Services: Societal Dependence on Natural
Ecosystems* (Daily 1997) and an article by Robert Costanza and colleagues on the
value of the world's ecosystem services and natural capital (Costanza et al. 1997).
The goal of these publications was to make explicit the contribution of ecosystems
to human well-being, and put an economic value on it (between 16 and 54 trillion
USD per year at the time), in order to make transparent the trade-offs involved in
any decision concerning the use of land and natural resources. This monetary
approach stirred a fierce debate, which is still ongoing, but eventually the concept of
ecosystem service met biodiversity conservation: first, in 2001, with the launch of
the Millennium Ecosystem Assessment (MA) by the United Nations Environment
Programme, and later, in 2007, with The Economics of Ecosystems and Biodiversity
(TEEB) initiative promoted by the German Government and the European
Commission. These programmes are focused, respectively, on the ecological and
economic aspects of ecosystem services, and are based on a utilitarian view of biodiversity (biodiversity must be preserved as an ecosystem service in itself, or as a

---

[4] With an argument intersecting that expounded by Alessandro Minelli, Chap. 11, in this volume.

component of the environment necessary for the maintenance of other ecosystem services), and on the implicit (and controversial) assumption that the protection of the ecosystem services leads to the protection of biodiversity (Mace et al. 2012).

15.2.2 *Ecosystem Services and Biodiversity: Epistemological and Ethical Troubles*

Biodiversity is considered a cultural service or an actual good (which might be marketable or not) when it provides non-material benefits to human beings. Wildlife, uncontaminated landscapes, totemic, charismatic and rare or endangered species have a particular appeal to human beings, because they respond to aesthetic, spiritual, religious, educational and recreational values. In these cases, people value the *diversity* of life as such—or some specific actualization of that diversity, as for instance charismatic species—and not some product or purported effect of biodiversity (e.g., variety of food or possibility to discover new drugs).[5] For all the other services, the relationship between biodiversity and human benefits is all but clear and needs to be examined on a case by case basis (Harrison et al. 2014). As a general rule, there is stronger evidence for the effects of biodiversity on ecosystems stability than on ecosystem services (Cardinale et al. 2012; Srivastava and Vellend 2005), and although it is generally agreed that biodiversity plays an insurance role, by potentially buffering ecosystems against environmental changes (Cottingham et al. 2001; Hooper et al. 2005; Loreau 2010a), data reviews and meta-analysis on the threefold relationship between biodiversity, ecosystem functioning, and ecosystem services are hampered by the lack of unified definitions and measures of biodiversity, and by the complexity and multi-faceted nature of each of the factors of the equation (Cardinale et al. 2012; Mace et al. 2012). Also, in many cases it is difficult to establish if the biodiversity effect is due to diversity as such (e.g., at the level of species, genes, or traits) or to other factors such as composition or biomass.

As mentioned above, within the ecosystem services approach, ecosystem services and biodiversity are often used as synonyms, thus implying that they are the same thing and that, by protecting one, we are automatically protecting the other (Costanza et al. 2017; TEEB 2010). On the contrary, within the conservationist perspective, biodiversity is an ecosystem service or a good *per se,* and as such it does not necessarily contribute to other ecosystem services and is potentially in conflict with them. Both positions have pitfalls. For what concerns the conservationist perspective, the main problem is that it is blind to the functional role of biodiversity, and often focuses on charismatic or endangered species. In so doing it loses sight of the greater variety of units, levels and scales at which biodiversity occurs, and perpetuates a static vision of life both at the species and ecosystem level. On the contrary, within the ecosystem services perspective, the functional role of biodiversity

---

[5] But for a problematisation of the relationship between biodiversity and cultural services see, for instance, Sarkar 2005, Cardinale et al. 2012.

is acknowledged, but in practice ecologists account for its contribution to the eco-system almost exclusively in terms of simple trophic structures and the related stocks and flows of energy, nutrients and biomass. This poses epistemological prob-lems related to the different aims, conceptual frameworks, and methodologies adopted in different scientific disciplines, where such problems call for theoretical and empirical solutions. Also, values of biodiversity other than its contribution to ecosystem functioning are not taken into account, thus posing an ethical problem (Mace et al. 2012).

The ethical criticism is the one most often leveraged against the ecosystem ser-vices approach (Reyers et al. 2012), and can be framed within a number of related debates: the controversy on the monetary nature of the concept of ecosystem service (e.g., McCauley 2006; Redford and Adams 2009); the debate about the instrumental *versus* intrinsic value of biodiversity (e.g., Norton 1986; Sarkar 2005; Maquire and Justus 2008; Justus et al. 2009); or the opposition between ecocentrism and anthro-pocentrism in environmental ethics (e.g., Singer 1975; Thompson and Barton 1994; Naess 1973). In this chapter, I let aside the ethical issues and focus on the epistemo-logical problems instead, trying to understand whether a stronger integration between ecology and evolutionary theory might make a difference in conservation planning within the ecosystem services approach.

15.2.3   Ecosystem Services and Biodiversity: An Ecologist's Perspective

For those who embrace the conservation perspective, there is a potential opposition between biodiversity and ecosystem services, and some authors see the ecosystem services approach as an unwarranted thwarting of the original mission of conserva-tion, namely, the protection of biodiversity or, more generally, nature, for its own sake (e.g., McCauley 2006; Redford and Adams 2009). From this perspective, the ecosystem services approach is detrimental to biodiversity conservation. However, if one tackles this criticism from an epistemological point of view, letting aside the controversy concerning the value of biodiversity, it becomes apparent that the endorsement of the concept of ecological service in many conservation policies has produced at least one major benefit for biodiversity science in that it has given spe-cial impulse to the study of the effects of biodiversity on ecosystem functioning in experimental and theoretical ecology (Loreau 2010). According to Loreau, this had relevant consequences for ecology both at the epistemological and disciplinary level. At the epistemological level, it has revived and reshaped the diversity-stability debate—that has run through ecosystem ecology since the 1950s (e.g., MacArthur 1955; May 1973; Pimm 1984)[6]—, and has given momentum to the study of the respective roles of individual-level and ecosystem-level selection in shaping ecosys-

---

[6] See Huneman, Chap. 13, in this volume, for a discussion of the notions of diversity used in the formulation and test of the stability hypothesis (biodiversity as an *explanans*).

tem properties—a controversial issue in both ecology and evolutionary biology (see Williams and Lenton 2007; Loreau 2010b). More importantly, it has changed the way ecosystem and community ecologists approach the study of biodiversity, giving prominence to the idea that biodiversity, especially species and genes diversity, is a driver of ecosystem functioning (Naeem 2002; Loreau 2010), and populations cannot be studied as homogeneous biomass pools in which individuals operate in identical ways to influence the nutrient and energy flows amongst the ecosystem compartments (Bassar et al. 2010).

At the disciplinary level, the need to better understand the effects of biodiversity on ecosystem functions at different spatial and temporal scales has made more evident and urgent the importance of integrating community ecology, ecosystem ecology and evolutionary biology (Loreau 2010, b).[7] Indeed, the development of the ecosystem services approach in environmental protection and biodiversity conservation has not only turned the study of the relationship between biodiversity and ecosystems into a pressing scientific matter, imposing a research agenda on ecologists (i.e., to understand the role and relevance of biodiversity for the delivery of ecosystem services). It has also implicitly indicated the scientific hypothesis to be tested, namely that biodiversity is necessary for ecosystem processes and that the loss of biodiversity hampers the functioning of ecosystems in the short and/or long term, thus affecting the provision of ecosystem services.

To answer the practical questions raised by the ecosystem services approach it is necessary to understand how ecosystems function and predict how they might change under a variety of environmental and anthropic pressures, such as climate change, habitat loss and degradation, overharvesting and diffusion of invasive exotic species. All these factors affect biodiversity as much as ecosystems as a whole. Loreau agrees with Mace and colleagues that current models of interaction between biodiversity and ecosystem functioning, based mostly on the modelling of evolutionary complex food webs, have several limitations. He stresses that important insights might come from theories such as ecosystem engineering (Jones et al. 1994, 1997; Wright and Jones 2006) and niche construction (Laland et al. 1999; Odling-Smee et al. 2003), which try to account for the ability of organisms to transform their habitat with relevant consequences both at the ecological and evolutionary level. In the last decade, there has been a surge of interest for eco-evolutionary theories (Whitham et al. 2006; Fussman et al. 2007), particularly in theoretical ecology (Kokko and Lopez-Sepulcre 2007). In what follows I present and discuss David Post and Eric Palkovacs' eco-evolutionary feedback (EEFB) theory, because it is an interesting example of ecological re-elaboration and clarification of the niche construction theory (henceforth NCT) originally formulated by Kevin Laland and John Odling-Smee, and also because Post and Palkovacs suggest that an integration of ecological and evolutionary theories would have relevant consequences not only for our understanding of ecosystem functioning, but also for biodiversity conservation.

---

[7] But see Huneman, Chap. 13, in this volume, for a criticism of this endeavour.

15.3   Eco-Evolutionary Feedback Theory

An eco-evolutionary feedback is "the cyclical interaction between ecology and evolution such that changes in ecological interactions drive evolutionary change in
organismal traits that, in turn, alter the form of ecological interactions, and so forth"
(Post and Palkovacs 2009). This description of the reciprocal causation between
ecological and evolutionary change clarifies the ecological relevance of NCT by
making a clear distinction between the process of niche construction, defined as
"the effect of an organism on its environment" (Post and Palkovacs 2009), and the
evolutionary feedbacks that occur in response to the environmental changes caused
by organisms. Niche construction *sensu stricto* (Post and Palkovacs 2009) includes
both active engineering and the effects caused by the by-products of biological process, while the evolutionary feedback can be the result of heritable traits change or
phenotypic plasticity. By explicitly separating the general process of EEFB into two
sub-processes (niche construction + evolutionary feedback), EEFB theory makes
clear that not all the biotic processes that shape the environment can cause subsequent evolution, because many factors can prevent the evolutionary feedback.
However, when the feedback occurs, it has important consequences at both the evolutionary and ecological level, because it can affect the direction of evolution and
alter the role of species in the ecosystem. It also highlights that both processes, even
when they do not occur together, have important ecological and evolutionary consequences, hence deserving in-depth study. Finally, unlike NCT, at least in its initial
version, EEFB allows for cases in which the recipient population of the modified
selective pressure can be different from the population that produced the environmental transformation in the first place (see Odling-Smee et al. 2013; Barker and
Odling-Smee 2014).

For an EEFB to occur, three conditions need be satisfied: (1) organisms must
have a phenotype that *strongly* impacts the environment, i.e., they must structure or
construct their niche (e.g., nutrients cycling and translocation, habitat construction
and modification, consumption)[8]; (2) the changes produced in the environment must
cause selection on a population and that this population has sufficient genetic capacity to evolve in response to changes in the environment; (3) the time-scales of the
ecological and evolutionary responses have to be congruent, i.e., the constructed
niche must persist for a duration that is sufficient to select the relevant traits (this
corresponds to the concept of ecological inheritance in NCT).

For what concerns (2) it should be noticed that, as in adaptive evolution more
generally, the evolutionary factors that determine whether a population will evolve
or go extinct are a combination of genetic factors (e.g., high levels of genetic variation are expected to favour evolutionary change); demographic factors (e.g.,

---

[8] Potentially, all organisms are niche constructors, because all organisms interact with the environment. However, as it will be explained below, a key factor for the identification of meaningful cases
of niche construction in the EEFB theory is the strength (magnitude and/or extent) of the interaction between an organism and the environment (which includes other organisms), and the spatial
and temporal scale of the effects of such interaction.

population size and genetic drift); and ecological factors (e.g., the rate of deforestation or the introduction of a toxic compound).

For what concerns (3), what counts as a *sufficient duration* will depend on the niche, as well as on the species and traits under consideration. In any case, there must be an overlap of ecological and evolutionary time: the constructed niche must persist long enough to produce evolutionary effects, and evolution must be fast enough to feed back on the constructed niche and further influence it. Since what matters is the congruence between ecological and evolutionary time, in principle evolution does not need to be rapid for EEFB to emerge. Slow niche construction, such as the oxygenation of earth's atmosphere by cyanobacteria, can create eco-evolutionary feedbacks as much as rapid evolution associated with rapid niche construction. However, the study of EEFB associated with rapid evolution has the advantage of being more easily amenable to empirical tests, and is more likely to be relevant in terms of biodiversity protection and ecosystem services conservation practices.

15.3.1 *EEFB and Contemporary Evolution: Three Empirical Cases*

The existence of rapid contemporary evolution, i.e., the evolution of heritable traits over a few generations (Stockwell et al. 2003; Jones et al. 2009),[9] is neither particularly controversial in ecology nor in evolutionary biology. What is controversial is the overall ecological and evolutionary relevance (prevalence and magnitude) of this phenomenon. As a matter of fact, in spite of the accumulation of studies that in the course of the last 40 years have shown that a strict distinction between ecological and evolutionary time is unwarranted, ecologists still tend to ignore potential effects of evolution on ecological interactions, because they assume that evolution occurs on a much slower time scale than ecological dynamics (Bassar et al. 2010). On the other hand, evolutionary biologists tend to ignore the action of organisms on their environment, because it is considered too weak and flimsy to significantly change selection pressures (Laland and Sterelny 2006). Eco-evolutionary theories challenge these entrenched views. In fact, there is growing evidence that contemporary evolution is a widespread phenomenon—which concerns many traits and many organisms from all kingdoms—and the evidence for potential cases of eco-evolutionary feedbacks is growing. Here I summarise three of the five empirical cases reviewed by Post and Palkovacs (2009): alewives' speciation caused by pat-

---

[9] Rapid evolution, contemporary evolution and microevolution are sometimes used as synonyms, and definitions vary (e.g., Thompson 1998, Kinnison and Hairston 2007, Ashley et al. 2003). Here I follow Post and Palkovacs 2009 and use contemporary evolution to refer to the overlap of ecological and evolutionary times, irrespectively of the actual duration of the process.

terns of migration, its influence on zooplankton communities, and the subsequent evolution of foraging traits; the effect of the life histories of Trinidad guppies on nutrient cycling and its potential feedback on male guppies' phenotype; the soil-mediated impact of *Populus* leaf tannins levels on the development of adapted roots.

15.3.1.1   Alewives and Zooplankton

Along North America East coast, the ecological isolation of lakes from the ocean has led to the phenotypic differentiation of alewife (*Alosa pseudoharengus*) land-locked populations that differ from the original anadromous population in feeding morphology and prey selectivity. Anadromous fishes migrate up rivers from the ocean to spawn and then go back to the open sea. In this case, the alewives only temporarily affect the community structure of lacustrine zooplankton (niche construction via predation, Post and Palkovacs 2009) before they go back to the ocean, thus the duration of the constructed niche is not long enough to cause an eco-evolutionary feedback. On the contrary, in the landlocked populations, intense year-round predation pressure eliminates large-bodied preys and produces a lacustrine zooplankton community of relatively low biomass of small-bodied zooplankton throughout the year (persistent constructed niche). This exerts a strong selection for traits related to foraging on small zooplankton, so that the landlocked population has developed smaller mouth gape and narrower spacing between gill rakes compared to the ancestral anadromous population (evolutionary feedback). In this case there is strong evidence for a complete EEFB.

15.3.1.2   Trinidad Guppies and Nutrients Cycling

Observations in the wild have shown that the life-histories (age and size at maturity) of Trinidad guppies (*Poecilia reticulata*) are affected by predation pressure. In high-predation environments, guppies reach maturity at an earlier age and smaller size, and they reproduce more frequently giving birth to smaller offsprings, with important effects for the population phenotype. Mesocosm experiments have shown that under conditions of equal biomass, populations characterised by a high number of small individuals (high-predation environment) drive higher nutrients flows compared to populations with fewer larger individuals (low-predation environment), thus increasing the rates of primary production, i.e., algal biomass (constructed niche). This, in turn, might influence further differentiation amongst guppies' populations, for instance, by influencing traits such as male colour patterns, which are under natural and sexual selection, and are sensitive to the levels of algae-derived carotenoids in the environment (potential eco-evolutionary feedback).

15.3.1.3 *Populus* and Soil Nutrients Levels

Poplar trees are foundation species whose chemical effects on leaf litter strongly influence community dynamics and ecosystem processes. Observational studies have shown that intraspecific variation in condensed tannin levels in poplar trees' leaves controls decomposition and nitrogen mineralisation rates, as well as the composition of the microbial community in the soil, thus creating a microhabitat (constructed niche). Since high concentrations of tannins inhibit nutrients release from leaves litter, poplar trees with high tannin levels will have to cope with low nutrients levels. According to EEFB theory, these trees should display some form of adaptation. Indeed, a strong positive correlation between leaf tannin levels and the development of finer roots has been observed, thus providing indirect evidence for eco-evolutionary feedback. However, ecological factors such as the presence of other plant species, herbivores and nutrients loading might disrupt or reduce the strength of the feedback by altering the ecology of the soil.

It is worth noticing that it is not always clear whether contemporary evolution is due to heritable traits or phenotypic plasticity. However, as remarked by Palkovacs et al. (2012), although such distinction is fundamental to our understanding of evolutionary and ecological processes, in the context of conservation biology it might be more important, and urgent, to link phenotypic change and ecosystem dynamics, regardless of the specific causes of change. Also, considering that plasticity itself is a hereditary trait that evolves and can direct future phenotypic change, it is not always useful to draw a thick line between plasticity and genetic change in terms of potential ecological causes and effects (Ghalambor et al. 2007; Palkovacs et al. 2012).[10] What is most relevant here is to highlight that the species more likely involved in EEFB are also the most relevant in terms of ecosystem functioning, because they strongly affect the community and the ecosystem where they live. They can be keystone, foundation, or dominant species, ecosystem engineers, or species that alter nutrient cycles through translocation or recycling.

15.3.2 *EEFB, Niche Construction, and Ecosystem Engineering*

What all these organisms have in common is that they are strong interactors.[11] To be a strong interactor, however, often depends on the ecological context: foundation species in one habitat might be rare in another, weak interactors in species-rich communities might have strong effects in species-poor communities, and species that move nutrients will have very different impacts in low- compared to high-nutrient environments (Post and Palcovaks 2009; Paine 1966; Menge et al. 1994).

---

[10] See also Minelli, Chap. 11, in this volume.

[11] For a detailed discussion of the differences between strong interactors, in particular between keystone species and ecosystem engineers see Boogert et al. 2006.

Thus, the ability of a species to construct a persistent niche often depends on the overall conditions of the ecosystem and the community, which means that it can vary in space and time. In turn, the eco-evolutionary feedback, with its potential to alter and respond to environmental selective pressure, can lead to the differentiation of a population whose ecological role is different from that of the original population, thus affecting community and ecosystem dynamics. Indeed, there might be instances in which the change of the traits of a species is at least as important as its presence/absence in terms of ecological effects. In the case of the alewives from North American coastal lakes, for example, there is evidence that the differentiation of the landlocked population has influenced the evolution of one of its preys, *Daphnia ambigua*, and this is likely to cause further effects on trophic cascades, because *Daphnia* is itself a strong interactor (a dominant grazer for zooplankton) (Palkovacs et al. 2012).

A main feature of EEFB theory is that it highlights the fact that organisms actively build their environment and that species, species traits, and species ecological impacts are dynamic and vary across space and time. A consequence of this is that within the research framework set by eco-evolutionary theories, the functional role of biodiversity in an ecosystem cannot be understood simply in terms of more or less complex trophic webs. This simplifying idealisation has been at the core of the success of ecosystem ecology in the study of terrestrial global biogeochemistry, but it has been increasingly called into question by ecologists themselves at least since the 1990s (Loreau 2010). In particular, the concept of ecosystem engineering introduced by Clive Jones and colleagues (Jones et al. 1994, 1997; Wright and Jones 2006), often considered the ecological counterpart of Laland and Odling-Smee's NCT, has shown that connectance webs that describe the processes driven by ecosystem engineers should be studied along with trophic webs, if we are to accurately model the interactions between communities and ecosystems. Importantly, these studies have shown that the laws of conservation of mass and energy, as well as the stoichiometry rules used to model trophic webs, cannot be used to predict the structure and outputs of ecosystem engineering networks, for which specific qualitative and quantitative models have been proposed (Jones et al. 1997; Boogert et al. 2006).

Ecosystem engineers are "organisms that directly or indirectly modulate the availability of resources (other than themselves) to other species, by causing physical state changes in biotic and abiotic materials. In so doing they modify, maintain and/or create habitats" (Jones et al. 1994). Within EEFB theory, they are seen as strong candidates for eco-evolutionary feedbacks, together with keystone species (species, usually predators, whose impact on their community or ecosystem is much larger than would be expected from their abundance), dominant species (species that outnumber their competitors in abundance or total biomass), and foundation species (species that strongly influence the structure of the community, e.g., by creating habitats). Accordingly, studying ecosystems from an EEFB theory perspective implies to parse strong ecological interactors according to a range of qualitative and quantitative models, e.g., strong *per capita* interactions that produce effects in the short term *vs.* weak but continuous *per capita* interactions that produce cumulative effects in the long term. Trophic webs, then, are but one of the interaction networks that compose the overall connectivity of the ecosystem. The other crucial ecological relationships that need attention are the non-trophic interaction webs described by ecosystem engineering theory, and the environmentally-mediated gene-associations (EMGAs) theorised by Odling-Smee and colleagues (Odling-Smee et al. 2003, 2013; Barker and Odling-Smee 2014), in a development of the original NCT, prompted by the insights provided by ecosystem engineering and eco-evolutionary theories.

15.3.3 EEFB and Environmentally-Mediated Gene-Associations

EMGAs are "indirect but specific connections between distinct genotypes mediated either by biotic or abiotic environmental components in the external environment [...]. They map sources of selection stemming from one population's genes onto genotypes in another population that evolve in response to those modified sources" (Odling-Smee et al. 2013). These *indirect* evolutionary interactions mediated by the environment emerge when the niche constructed by a population—via its physiological processes as well as active engineering—influences the selective pressure acting on the same population or, more often, on a different population of different species. For example, in the case of Trinidad guppies, predators, through differential predation pressure, can influence guppy populations' life histories, leading to the differentiations of populations of larger or smaller guppies, characterised by different rates of excretion that determine differential inorganic nutrients distribution. This, in turn, affects algal growth, which has the potential to feed back on the selection of male guppy colour patterns through the concentration of carotenoids released by algae in the environment.

The idea of EMGAs helps formalise the causal chain of EEFB in genetic terms, and can be used to visualise the ramifications of evolutionary and ecological effects deriving from niche construction via biotic or abiotic mediations. In its original form, it gives epistemic priority to the genetic component within the EEFB's causal chain, but in those cases in which the niche construction is underpinned by non-heritable variation, environmentally-mediated *genotypic*-associations are replaced by environmentally-mediated *phenotypic*-associations (EMPAs), thus emphasising that the phenotype should not be thought of as the mere epiphenomenon of genetic information, but as the dynamic result of the combination of heritable variation with a number of non-heritable factors, such as plasticity, epigenetics and population structure (Odling-Smee et al. 2013).

It follows that, in order to respond to the requirements of EEFB theory, the study of ecosystem processes and functioning should be articulated along two interrelated axes, which force ecosystem ecology to revise its operational simplifying idealisations. On the one hand, the study of the sub-process of niche construction requires the development of ecosystem models that account for high degrees of connectance at the different scales of the ecosystem, integrating trophic and competitive webs with more complex interaction webs, as well as EMGAs or EMPAs; on the other hand, the study of the sub-process of evolutionary feedbacks needs to be carried out taking into account both genetic and non-heritable phenotypic variation, because both can be sources of functional evolution and adaptation. Accordingly, functional diversity must be understood as a dynamic epiphenomenon that can potentially emerge from both genetic and non-genetic factors that need to be studied on a case by case basis.

All in all, what emerges from EEFB theory is a highly dynamic picture of ecosystems, populations and communities, in which the structure of biodiversity—used here as a shorthand for diversity at the level of species, genes, traits, communities, etc.—can vary more easily than both ecologists and evolutionary biologists are prone to believe, and where the causal chain of change does not go exclusively from the environment to the organism (ecological change as a cause of trait change), but can go from the organism to the environment (trait change as a cause of ecological change). In the next section, I explore the potential consequences of this shift of perspective for conservation biology.

15.4 Eco-Evolutionary Feedback Theory: Some Consequences for Biodiversity Conservation

The study of EEFB pushes ecologists to recognise that contemporary evolution creates phenotypic differences that can alter the role of a species in a community or ecosystem at ecological time-scales. This implies that evolution can no longer be considered mere background noise in the study of ecosystem dynamics, and extant and potential novel biodiversity become a fundamental component of the study of ecosystem dynamics. For Post and Palkovacs: "the study of eco-evolutionary feedbacks focuses attention on the bidirectional interactions that unify ecology and evolution, and highlights the importance of conserving both ecological and evolutionary diversity in nature" (Post and Palkovacs 2009). But how, exactly, could EEFB theory guide biodiversity conservation? As referred to in Sect. 15.1, a criticism leveraged by conservation biologists to ecologists within the ecosystem services approach is that they account for biodiversity's contribution to ecosystem functions almost exclusively in terms of simple trophic structures (Mace et al. 2012). What kind of instruments does EEFB theory offer to tackle this issue?

15.4.1 Ecosystem Engineers First?

Considering that EEFB theory has many points in common with the ecosystem engineering theory, some important insights about the impact of eco-evolutionary theories on biodiversity conservation can be found in Crain and Bertness 2006 and in Boogert et al. 2006. For these authors, ecosystem engineers should be the primary targets of biodiversity conservation policies, because they shape habitats and ecosystems, with all their related species and functions. Since ecosystem engineers are responsible for a much higher and more complex level of inter-species connectance than the trophic webs generated by other organisms, the loss of ecosystem engineers is more likely to have far reaching negative consequences on both communities and whole ecosystems (Crain and Bertness 2006. See also Jones et al. 1997; Wright and Jones 2006). Although species that are ecosystem engineers under certain circumstances may not be so under others, it is possible to identify fundamental engineering roles in ecosystems, independently of the specific species involved. Accordingly, to grant stability to ecosystem structure and functioning, conservation policies should focus on protecting the activity of key engineers, rather than the species composition of an ecosystem (Boogert et al. 2006; see also Odling-Smee et al. 2003). This is a classical argument in favour of the preservation of functional diversity rather than species diversity, and is usually criticised for being too narrow a criterion for selecting the aspects of biodiversity worth protection (Mace et al. 2012). To preserve ecosystem functioning, in fact, we do not need to protect all the species that perform a given function and their genetic variability. For instance, we do not need to protect all the species of trees in a forest, and their intraspecific variation, to ensure biomass production, oxygen emission, and CO_2 sequestration. From this perspective, the most efficient course of action would be to select the species that better perform the function of interest, and focus our conservation efforts on them. This approach is likely to leave aside rare species, which represent a primary target for conservation biology, because their functional role on an ecosystem is often negligible. In this respect, not only do aspects of biodiversity not related to ecological functions become irrelevant, but the replacement of ecosystem engineering species using artificial solutions becomes an acceptable option (e.g., replacing of caterpillars by artificially created leaf ties, see Lill and Marquis 2003). Here, at least in principle, the choice to favour technological solutions over biodiversity conservation will be constrained by considerations of efficacy and cost-effectiveness (Boogert et al. 2006), rather than by an *a priori* obligation to avoid species extinction, or a precautionary approach whereby a species (or a genome) that has no particular functional import under the present conditions might become relevant under different conditions, because of ecological changes or because our knowledge of the benefits we obtain from that particular species/genome changes (see Maclaurin and Sterelny 2008; Sarkar 2005). Thus, although there are compelling reasons for choosing ecosystem engineers as targets of biodiversity conservation, this choice must be further qualified and refined.

15.4.2 Genetic Diversity: Better Safe than Sorry

Niche construction (of which ecosystem engineering is just one possible case) is only one half of the EEFB process. The other is evolutionary feedback. To the extent that evolutionary feedbacks have the potential to produce relevant ecological effects, they should be taken into account in conservation policies aimed at preserving ecosystem functions. Since one of the conditions for EEFB is that the population(s) affected by the constructed niche must have sufficient genetic capacity to evolve in response to new selective pressures before going extinct, it follows that it is important to preserve not only functional diversity, but also genetic diversity (that might include phenotypic plasticity), because this ensures that niche constructing species, or other species potentially affected by the constructed niche, will maintain their ability to respond to environmental modifications. In this respect, it should be noted that trait change *per se* is not a guarantee of ecosystem stability, because phenotypic variation can be both a driver of or a buffer against ecological change. In the empirical cases described in Sect. 15.3.1, in fact, the putative evolutionary feedback works clearly as a stabiliser of functions only in the case of poplar trees, while in the cases of the alewives and Trinidad guppies the evolutionary feedback potentially causes a cascade of changes in the community structures whose consequences in terms of ecosystem functions need further clarification. This only makes the need to improve our understanding of eco-evolutionary interactions more compelling, in order to be able to predict when they could buffer the ecosystem and when they would magnify potential functional disruptions. Sweeping generalisations are not warranted in this relatively recent domain of inquiry, but there is evidence that contemporary evolution is most common, although less evident, when it counteracts phenotypic changes caused by environmental pressure, thus buffering ecosystem functions (Ellner et al. 2011; Palkovacs et al. 2012). Preserving the genetic diversity that feeds contemporary evolution, then, seems a safe bet.

Without entering into the debate on what genetic diversity exactly is, how to measure it, and what to do to preserve it (see Mace and Purvis 2008 for a list of problems in this field), we can say that, by providing a clear and well-structured theoretical framework for the empirical study of the reciprocal interaction between evolutionary and ecological processes, EEFB theory offers decisive evidence for the necessity of keeping into account evolutionary dynamics in the study of ecosystem functioning. Accordingly, it provides arguments to support the importance of "evolutionary-enlightened management" in biodiversity conservation (Ashley et al. 2003). In fact, whether the eco-evolutionary feedbacks magnify ecological change or buffer against it, they must be taken into account if we are to preserve ecosystems functioning.

15.4.3   EEFB Theory and Evolutionary-Enlightened Management

For the proponents of evolutionary approaches to biodiversity protection, conservation policies are hampered by the misplaced idea that while human disturbance is very fast, adaptation is a very slow process, thus irrelevant to conservation planning, whose temporal horizon seldom exceeds a few decades (Mace and Purvis 2008). Typological thinking concerning both species and ecosystems is another hindrance to evolutionary-enlightened management, since it promotes the idea that evolutionary change has relevant consequences at an ecological and human time-scale only when it concerns organisms with short generation time (e.g., microorganisms).

Consequently, it is argued, its effects on the whole of biodiversity are negligible (Ashley et al. 2003; Santamaria and Mendez 2012). Mary Ashley and colleagues also remark that, although in conservation planning it is theoretically acknowledged that species respond to change both ecologically and evolutionarily, in practice the importance of evolutionary responses is often neglected. For instance, research models on potential impacts of rising temperature and CO_2 concentrations generally make predictions concerning possible ecological adaptations based on the present ecologies of extant species, without taking into account evolutionary factors such as climate adaptation and the potential disruption to gene flow caused by climate change. Similarly, conservation approaches based on population viability analysis are based on models that assume that the life histories and demographic characteristics of a species are fixed (Ashley et al. 2003). Still, as seen in the example of the Trinidad guppies, environmental factors can significantly affect life histories, with relevant consequences for the structure of a population. This can in turn produce changes in the environment, e.g. in the recycling of nutrients, creating the conditions for further evolutionary feedbacks.

Rapid contemporary evolution is the main preoccupation of evolutionary-minded conservationists, not least because anthropic drivers of rapid evolution, such as habitat loss and degradation, overharvesting, and the introduction of exotic species, are also the factors that have led to the current extinction crisis (Stockwell et al. 2003; Palkovacs et al. 2012). EEFB theory reinforces this preoccupation because it draws on the evidence that rapid contemporary evolution is a widespread phenomenon. At the same time, one of its theoretical tenets is that eco-evolutionary feedbacks can occur at any timescale, thus highlighting that just as evolutionary factors must be taken into account not only in the long, but also in the short term, ecological effects of evolutionary change might become salient over long timescales. This happens, for instance, when a newly evolved trait constructs a niche whose effects slowly accumulate over time, because it has little *per capita* impact or because external factors intervene to dissipate or swamp the niche. Thus, the effects of EEFB can be time-lagged (Odling-Smee et al. 2013), and this makes predictions more complex, thus more prone to error, but also more realistic.

The implementation of evolutionary-enlightened management for biodiversity conservation would imply the development of research programmes that incorporate evolution into applied ecology and resource management; the assessment of populations' short-term evolutionary potential using direct measures of genetic variation rather than the proxy of neutral molecular variation; and the use of quantitative genetics to assess the genetic variability of traits that are likely to be under selective pressure in hypothetical scenarios (Ashley et al. 2003). Ecological and evolutionary interactions are extremely complex and it is very hard to create workable predicting models. EEFB theory *per se* does not provide a direct answer to this problem, but offers a theoretical framework that can favour the development of such models. Post and Palkovacs' simple move of refining the NCT by splitting the EEFB into two well defined sub-processes allows to break down intricate eco-evolutionary pathways into more tractable components, which can be analysed at different spatial and temporal scales (from long-term whole-ecosystem observation to short-term, small-scale experiments). Subsequently, the general picture can be reconstructed by retracing the network of interactions, their strength and their variation over time (see Odling-Smee et al. 2013). As pointed out in Barker and Odling-Smee (2014), in order to be able to make predictions about the evolution of whole ecosystems and of their components, we need to bring together theories that are general and realistic enough to afford a "local theoretical unification" with precise and realistic models that describe the details of particular complex systems, providing "explanatory concrete integration" (Mitchell 2002). Theories such as EEFB are good candidates for making this synthesis, because they favour the integration of ecosystem ecology, population ecology, and evolutionary biology, and their respective methodological frameworks. If EEFB theory proved successful, then, we would be able to overcome the problem of having too simplified an account of ecosystem functioning and it would be possible to clarify the role of functional diversity within ecosystem processes.

15.5   Conclusions

Since the late 1990s, the development of the concept of ecosystem service for conservation policies has given new momentum to the study of the effects of biodiversity on ecosystem functioning in experimental and theoretical ecology, revitalising the traditional diversity-stability debate and fostering the development of ecosystem evolution theories.

EEFB theory emphasises the active role of organisms in shaping their environment and supports the idea that contemporary evolution is a common and widespread phenomenon. This means that species, their traits, and their ecological impacts are dynamic and vary across space and time. As a consequence, the functional contribution of biodiversity to ecosystem processes cannot be understood simply in terms of mass and energy conservation and stoichiometry rules for trophic webs, but must include, at least, the more elaborated connectance webs proposed by ecosystem engineering theory, and models of environmentally-mediated gene or phenotype associations proposed in recent developments of the NCT. Also, since contemporary evolution can be either a source of ecological change (potential disruption of ecosystem functions) or a buffer against change (preservation of ecosystem functions), in order to make predictions on the evolution of ecosystems and their capacity to sustain ecosystem services, we need to better understand eco-evolutionary interactions from the population to the whole-ecosystem level. On the whole, EEFB theory provides a non-typological image of both species and ecosystem, and challenges static visions in both ecology and evolutionary biology. On the one hand, it defies the idea that evolution is too slow to be relevant in the modelling of ecosystem processes; on the other hand, it undermines the idea that the action of organisms on their environment is too ephemeral to direct selective pressures. All in all, this calls for an evolutionary-enlightened management of biodiversity.

Ultimately, by emphasising the fact that organisms are active agents of ecological and evolutionary change rather than passive objects of selection, EEFB theory causes a shift of perspective on the role of biodiversity in the transformation of ecosystems. In fact, if "organisms and their local environments [are] integrated systems that evolve together" (Barker and Odling-Smee 2014), then species and genetic diversity are at least as important as functional diversity for the evolution and future functioning of an ecosystem. Now, to be able to make predictions about the potential evolution of ecosystems is a fundamental feature of the ecosystem services approach. By definition, what matters the most within the ecosystem services approach is to preserve functional ecosystems, so that humans can receive benefits from them. Accordingly, biodiversity is valued for what it can deliver in terms of ecological functions (with the sole exception of cultural services, where biodiversity can be relevant for its existence value). But in a scenario of locally co-evolving organisms and ecosystems, functions can be preserved only if we can preserve the possibility of organismal change. This implies to protect species and genetic diversity together with functional diversity. While the latter grants ecosystem functioning in the present, the former influences the ability of the ecosystem to continue to function under changing conditions, which can be generated in the long as well as the short term by the internal dynamics of eco-evolutionary change or by external ecological pressures, often of anthropic origin. In ecosystem services parlance, this increases the insurance value of biodiversity. Importantly, the idea of evolving species in evolving ecosystems defies static and typological thinking in ecosystem services policies as much as in traditional biodiversity conservation, thus fostering dynamic approaches and long-term planning.

Acknowledgments This work was funded by the Fundação para a Ciência e a Tecnologia through a postdoctoral grant within the R&D project Biodecon (PTDC/IVC-HFC/1817/2014).

References

Ashley, M. V., Willson, M. F., Pergams, O. R., O'Dowd, D. J., Gende, S. M., & Brown, J. S. (2003). Evolutionarily enlightened management. *Biological Conservation, 111*(2), 115–123. https://doi.org/10.1016/S0006-3207(02)00279-3.

Barker, G., & Odling-Smee, F. J. (2014). Integrating ecology and evolution: Niche construction and ecological engineering. In G. Barker, E. Desjardins, & T. Pearce (Eds.), *Entangled life. Organism and environment in the biological and social sciences* (pp. 187–211). Dordrecht: Springer. https://doi.org/10.1007/978-94-007-7067-6.

Bassar, R. D., Marshall, M. C., López-Sepulcre, A., Zandonà, E., Auer, S. K., Travis, J., Pringle, C. M., Flecker, A. S., Thomas, S. A., Fraserg, D. F., & Reznicka, D. N. (2010). Local adaptation in Trinidadian guppies alters ecosystem processes. *Proceedings of the National Academy of Sciences, 107*(8), 3616–3621. www.pnas.org/cgi/doi/10.1073/pnas.0908023107.

Boogert, N. J., Paterson, D. M., & Laland, K. N. (2006). The implications of niche construction and ecosystem engineering for conservation biology. *AIBS Bulletin, 56*(7), 570–578. https://doi.org/10.1641/0006-3568(2006)56[570:TIONCA]2.0.CO;2.

Cardinale, B. J., Duffy, J. E., Gonzalez, A., Hooper, D. U., Perrings, C., Venail, P., Narwani, A., Mace, G. M., Tilman, D., Wardle, D. A., Kinzig, A. P., Daily, G. C., Loreau, M., Grace, J. B., Larigauderie, A., Srivastava, D. S., & Naeem, S. (2012). Biodiversity loss and its impact on humanity. *Nature, 486*, 59–67. https://doi.org/10.1038/nature11148.

Costanza, R., d'Arge, R., de Groot, R., Farber, S., Grasso, M., Hannon, B., Limburg, K., Naeem, S., O'Neill, R. V., Paruelo, J., Raskin, R. G., Sutton, P., & van den Belt, M. (1997). The value of the world's ecosystem services and natural capital. *Nature, 387*(6630), 253–260.

Costanza, R., de Groot, R., Braat, L., Kubiszewski, I., Fioramonti, L., Sutton, P., Farber, S., & Grasso, M. (2017). Twenty years of ecosystem services: How far have we come and how far do we still need to go? *Ecosystem Services, 28*, 1–16. https://doi.org/10.1016/j.ecoser.2017.09.008.

Cottingham, K. L., Brown, B. L., & Lennon, J. T. (2001). Biodiversity may regulate the temporal variability of ecological systems. *Ecology Letters, 4*(1), 72–85. https://doi.org/10.1046/j.1461-0248.2001.00189.x.

Crain, C. M., & Bertness, M. D. (2006). Ecosystem engineering across environmental gradients: Implications for conservation and management. *AIBS Bulletin, 56*(3), 211–218. https://doi.org/10.1641/0006-3568(2006)056[0211:EEAEGI]2.0.CO;2.

Daily, G. C. (1997). *Nature's services: Societal dependence on natural ecosystems.* Washington, DC: Island Press.

Ellner, S. P., Geber, M. A., & Hairston, N. G. (2011). Does rapid evolution matter? Measuring the rate of contemporary evolution and its impacts on ecological dynamics. *Ecology Letters, 14*(6), 603–614. https://doi.org/10.1111/j.1461-0248.2011.01616.x.

Fussmann, G. F., Loreau, M., & Abrams, P. A. (2007). Eco- evolutionary dynamics of communities and ecosystems. *Functional Ecology, 21*, 465–477. https://doi.org/10.1111/j.1365-2435.2007.01275.x.

Ghalambor, C. K., McKay, J. K., Carroll, S. P., & Reznick, D. N. (2007). Adaptive versus non-adaptive phenotypic plasticity and the potential for contemporary adaptation in new environments. *Functional Ecology, 21*(3), 394–407. https://doi.org/10.1111/j.1365-2435.2007.01283.x.

Harrison, P. A., Berry, P. M., Simpson, G., Haslett, J. R., Blicharska, M., Bucur, M., Dunford, R., Egoh, B., Garcia-Llorente, M., Geamănă, N., & Geertsema, W. (2014). Linkages between biodiversity attributes and ecosystem services: A systematic review. *Ecosystem Services, 9*, 191–203. https://doi.org/10.1016/j.ecoser.2014.05.006.

Hooper, D. U., Chapin, F. S., Ewel, J. J., Hector, A., Inchausti, P., Lavorel, S., Lawton, J. H., Lodge, D. M., Loreau, M., Naeem, S., & SchmidT, B. (2005). Effects of biodiversity on ecosystem functioning: A consensus of current knowledge. *Ecological Monographs, 75*(1), 3–35.

Jones, C. G., Lawton, J. H., & Shachak, M. (1994). Organisms as ecosystem engineers. *Oikos, 69*, 373–386. http://www.jstor.org/stable/3545850.

Jones, C. G., Lawton, J. H., & Shachak, M. (1997). Positive and negative effects of organisms as physical ecosystem engineers. *Ecology, 78*(7), 1946–1957. https://doi.org/10.1890/0012-9658(1997)078[1946:PANEOO]2.0.CO;2.

Jones, L. E., et al. (2009). Rapid contemporary evolution and clonal food web dynamics. *Philosophical Transactions of the Royal Society B, 364*, 1579–1591. https://doi.org/10.1098/rstb.2009.0004.

Justus, J., Colyvan, M., Regan, H., & Maguire, L. (2009). Buying into conservation: Intrinsic versus instrumental value. *Trends in Ecology and Evolution, 24*(4), 187–191. https://doi.org/10.1016/j.tree.2008.11.011.

Kinnison, M. T., & Hairston, N. G. (2007). Eco-evolutionary conservation biology: Contemporary evolution and the dynamics of persistence. *Functional Ecology, 21*(3), 444–454. https://doi.org/10.1111/j.1365-2435.2007.01278.x.

Kokko, H., & Lopez-Sepulcre, A. (2007). The ecogenetic link between demography and evolution: Can we bridge the gap between theory and data? *Ecology Letters, 10*, 773–782. https://doi.org/10.1111/j.1461-0248.2007.01086.x.

Laland, K. N., Odling-Smee, F. J., & Feldman, M. W. (1999). Evolutionary consequences of niche construction and their implications for ecology. *Proceedings of the National Academy of Sciences, 96*(18), 10242–10,247. https://doi.org/10.1073/pnas.96.18.10242.

Laland, K. N., & Sterelny, K. (2006). Perspective: Seven reasons (not) to neglect niche construction. *Evolution, 60*(9), 1751–1762. https://doi.org/10.1111/j.0014-3820.2006.tb00520.x.

Lill, J. T., & Marquis, R. J. (2003). Ecosystem engineering by caterpillars increases insect herbivore diversity on white oak. *Ecology, 84*(3), 682–690. https://www.jstor.org/stable/3107862.

Loreau, M. (2010). Linking biodiversity and ecosystems: Towards a unifying ecological theory. *Philosophical Transactions of the Royal Society, B: Biological Sciences, 365,* 49–60. https://doi.org/10.1098/rstb.2009.0155.

Loreau, M. (2010a). *The challenges of biodiversity science.* Oldendorf/Luhe: International Ecology Institute.

Loreau, M. (2010b). *From populations to ecosystems: Theoretical foundations for a new ecological synthesis (MPB-46).* Princeton/Woodstock: Princeton University Press.

MA, Millennium Ecosystem Assessment. (2003). *Ecosystems and human well-being: A framework for assessment.* Washington, DC: Island Press.

MA, Millennium Ecosystem Assessment. (2005). *Ecosystems and human well-being: Synthesis.* Washington, DC: Island Press.

Maclaurin, J., & Sterelny, K. (2008). *What is biodiversity?* Chicago: University of Chicago Press.

May, R. M. (1973). *Stability and complexity in model ecosystems. Monographs in population biology.* Princeton: Princeton University Press.

MacArthur, R. H. (1955). Fluctuations of animal populations and a measure of community stability. *Ecology, 36,* 533–535. https://doi.org/10.2307/1929601.

Mace, G. M., & Purvis, A. (2008). Evolutionary biology and practical conservation: Bridging a widening gap. *Molecular Ecology, 17,* 9–19. https://doi.org/10.1111/j.1365-294X.2007.03455.x.

Mace, G. M., Norris, K., & Fitter, A. H. (2012). Biodiversity and ecosystem services: A multilayered relationship. *Trends in Ecology & Evolution, 27*(1), 19–26. https://doi.org/10.1016/j.tree.2011.08.006.

Maquire, L. A., & Justus, J. (2008). Why intrinsic value is a poor basis for conservation decisions. *BioScience, 58,* 910–911.

Menge, B. A., Berlow, E. L., Blanchette, C. A., Navarrete, S. A., & Yamada, S. B. (1994). The keystone species concept: Variation in interaction strength in a rocky intertidal habitat. *Ecology Monographs, 64,* 249–286. https://doi.org/10.2307/2937163.

McCauley, D. J. (2006). Selling out on nature. *Nature, 443,* 27–28.

Mitchell, S. D. (2002). Integrative pluralism. *Biology and Philosophy, 17*(1), 55–70.

Naeem, S. (2002). Ecosystem consequences of biodiversity loss: The evolution of a paradigm. *Ecology, 83,* 1537–1552. https://doi.org/10.1890/0012-9658(2002)083[1537:ECOBLT]2.0.CO;2.

Naess, A. (1973). The shallow and the deep, long range ecology movement. A summary. *Inquiry, 16,* 95–100.

Norton, B. G. (1986). *The preservation of species: The value of biological diversity.* Princeton: Princeton University Press.

Odling-Smee, F. J., Laland, K. N., & Feldman, M. W. (2003). Niche construction: The neglected process in evolution. In *Monographs in population biology.* Princeton: Princeton University Press.

Odling-Smee, J. F., Erwin, D. H., Palkovacs, E. P., Feldman, M. W., & Laland, K. N. (2013). Niche construction theory: A practical guide for ecologists. *The Quarterly Review of Biology, 88*(1), 3–28.

Paine, R. T. (1966). Food web complexity and species diversity. *Am. Nat. 100,* 65–75. https://doi.org/10.1086/282400.

Palkovacs, E. P., Kinnison, M. T., Correa, C., Dalton, C. M., & Hendry, A. P. (2012). Fates beyond traits: Ecological consequences of human-induced trait change. *Evolutionary Applications, 5*(2), 183–191. https://doi.org/10.1111/j.1752-4571.2011.00212.x.

Pimm, S. L. (1984). The complexity and stability of ecosystems. *Nature, 307,* 321–326. https://doi.org/10.1038/307321a0.

Post, D. M., & Palkovacs, E. P. (2009). Eco-evolutionary feedbacks in community and ecosystem ecology: Interactions between the ecological theatre and the evolutionary play. *Philosophical Transactions of the Royal Society B, 364,* 1629–1640. https://doi.org/10.1098/rstb.2009.0012.

Redford, K. H., & Adams, W. M. (2009). Payment for ecosystem services and the challenge of saving nature. *Conservation Biology, 23,* 785–787. https://doi.org/10.1111/j.1523-1739.2009.01271.x.

Reyers, B., Polasky, S., Tallis, H., Mooney, H. A., & Larigauderie, A. (2012). Finding common ground for biodiversity and ecosystem services. *BioScience, 62*(5), 503–507. https://doi. org/10.1525/bio.2012.62.5.12.

Santamaría, L., & Mendez, P. F. (2012). Evolution in biodiversity policy–current gaps and future needs. *Evolutionary Applications, 5*(2), 202–218. https://doi. org/10.1111/j.1752-4571.2011.00229.x.

Sarkar, S. (2005). *Biodiversity and environmental philosophy: An introduction*. New York: Cambridge University Press.

Singer, P. (1975). *Animal liberation: A new ethics for our treatment of animals*. New York: Random House.

Srivastava, D. S., & Vellend, M. (2005). Biodiversity-ecosystem function research: Is it relevant to conservation? *Annual Review of Ecology, Evolution, and Systematics, 36*, 267–294. https://doi. org/10.1146/annurev.ecolsys.36.102003.152636.

Stockwell, C. A., Hendry, A. P., & Kinnison, M. T. (2003). Contemporary evolution meets conservation biology. *Trends in Ecology & Evolution, 18*, 94–101. https://doi.org/10.1016/ S0169-5347(02)00044-7.

TEEB. (2010). *The economics of ecosystems and biodiversity: Ecological and economic foundations*. Ed. P. Kumar. London/Washington, DC: Earthscan.

Thompson, S. C. G., & Barton, M. A. (1994). Ecocentric and anthropocentric attitudes toward the environment. *Journal of Environmental Psychology, 14*(2), 149–157. https://doi.org/10.1016/ S0272-4944(05)80168-9.

Thompson, J. N. (1998). Rapid evolution as an ecological process. *Trends in Ecology & Evolution, 13*(8), 329–332. https://doi.org/10.1016/S0169-5347(98)01378-0.

Westman, W. E. (1977). How much are nature's services worth? *Science, 197*(4307), 960–964.

Whitham, T. G., Bailey, J. K., Schweitzer, J. A., Shuster, S. M., Bangert, R. K., LeRoy, C. J., Lonsdorf, E. V., Allan, G. J., DiFazio, S. P., Potts, B. M., & Fischer, D. G. (2006). A framework for community and ecosystem genetics: From genes to ecosystems. *Nature Reviews Genetics, 7*, 510–523. https://doi.org/10.1038/nrg1877.

Williams, H. T. P., & Lenton, T. M. (2007). Artificial selection of simulated microbial ecosystems. *Proceedings of the National Academy of Sciences, 104*, 8918–8923. https://doi.org/10.1073/ pnas.0610038104.

Wright, J. P., & Jones, C. J. (2006). The concept of organisms as ecosystem engineers ten years on: Progress, limitations, and challenges. *BioScience, 56*(3), 203–209. https://doi. org/10.1641/0006-3568(2006)056[0203:TCOOAE]2.0.CO;2.

Part III
Scientific and Policy Perspectives on Biodiversity Conservation

16

Understanding Biodiversity: Issues and Challenges

Georg Toepfer

> *The complexity of the biodiversity concept does not only mirror the natural world it supposedly represents; it is that plus the complexity of human interactions with the natural world, the inextricable skein of our values and its value, of our inability to separate our concept of a thing from the thing itself. Don't know what biodiversity is? You can't.*
>
> David Takacs (1996, p. 341)

Abstract The impressive success of the concept 'biodiversity' in the last decades, in particular in the arena of politics, is in a large part due to its power to amalgamate facts and values: the fact that living beings show variety on every level of their existence, and the assumed values that are associated with this variety. These values are far from obvious or objective, they are rather normatively prescribed. They are already at work in the process of selecting the level of analysis, e.g. the level of genes, species, or ecosystems. The concept thus ties together many different discourses from the fields of biology and bioethics, aesthetics and economy, law and global justice. One important consequence of the concept's integrative power is the impossibility of its general definition. Just as 'life', 'time' or 'world' the word is an "absolute metaphor" or "non-concept" in the sense of Hans Blumenberg: it cannot have a fixed meaning just because it mediates between various contexts and disciplines. Any attempt to define 'biodiversity' in general terms is thus futile and does not capture the role it fulfills in contemporary discourse. Rather than trying to define the concept, reconstructing the interaction of its various contexts is a more promising approach. These include, besides the obvious reference to biology and nature conservation, ancient narratives about divine creation, paradise and Noah's ark as

G. Toepfer (✉)
Leibniz-Zentrum für Literatur- und Kulturforschung, Berlin, Germany
e-mail: toepfer@zfl-berlin.org

well as political ideas of pluralism, egalitarianism and non-hierarchical representation of individuals, or the values of market economy. In order to understand the current success and discursive role of the concept, I will analyze some of the underlying ideas, especially with respect to the representation of biodiversity in images.

16.1 The Integrative Power of 'Biodiversity'

In order to understand why 'biodiversity' has become such a successful term in the last decades it is necessary to look beyond biology to the broader socio-cultural and political contexts in which diversity became an important issue. From a merely biological point of view or from the standpoint of conservation biology it is not obvious that diversity should be more important than other abstract properties of biological systems such as stability, resilience or wilderness; or, to be more practical, the protection of one particularly endangered species or of an ecosystem. For the purpose of understanding why we now live in the "decade of biodiversity", as declared by the United Nations in 2010, it is not enough to point at increased biological knowledge in the modern age or at the crisis of mass extinction. It is necessary to focus on cultural and political values present in scientific issues—or at least in the public understanding of scientific issues.

The parallels between cultural and biological diversity are underlined by two conventions of the United Nations (Heyd 2010): The Rio "Convention on Biological Diversity" of 1992 explicitly calls for the protection and use of "biological resources in accordance with traditional cultural practices" (UNEP 1992). And the 2001 UNESCO "Declaration on Cultural Diversity" claims that "[a]s a source of exchange, innovation and creativity, cultural diversity is as necessary for humankind as biodiversity is for nature" (UNESCO 2001).

At least on a rhetorical level, 'diversity' functions as a versatile concept bringing together diverse fields. It can be linked to *economic* considerations: to plants and animals as entities providing "ecosystem services" regarding the supply of food, fibres or pharmaceuticals, or for the regulation of climate, water balance, etc. (Millennium Ecosystem Assessment 2005). At the same time, the term maintains its strong non-instrumental *ethical* dimension: it expresses a non-anthropocentric value of plants and animals. With this in mind the Holy Father, Pope Francis, in his 2015 encyclical *Laudato si'*, referred to biodiversity and ascribed intrinsic value to non-human creatures (Pope Francis 2015, no. 118; 140). Finally, 'biodiversity' has the dignity of a scientific term, as it seems to refer to something objective and measurable.

'Biodiversity' obviously forms an efficient basis for the integration of heterogeneous discourses and their public communication. Yet, by putting many things together—ethics, religion, science, business and politics—the term has an undifferentiating effect. On a political level this effect has also been welcomed because to be politically effective a term should not only describe a natural state of affairs but declare it as an important and good thing.

The main strength of the term seems to be that it does not take sides in fundamental ideological dichotomies such as scientific/emotional, profane/sacred or utilitarian/ intrinsic value. It remains neutral and thus can be used in either position. And, surprisingly, considering its integrative power and ideological neutrality, 'biodiversity' seems not to be abstract: It refers primarily to concrete individuals and species – well-liked species for the most part, the so-called "charismatic megafauna" such as polar bears, lions and elephants. These tangible references make 'biodiversity' a much more attractive concept, than, for example, 'stability', 'ecosystem services' or 'sustainability'. This suggestive concreteness is, of course, a huge advantage in the communication with the general public; it is a potent instrument for connecting nature and people (Díaz et al. 2015).

In addition to its integrative function and its concreteness, 'biodiversity' fits very well into our pluralistic present because the concept renounces an overarching, universally valid (world) order and expresses a de-hierarchization and pluralization of perspectives. It refers to the heterogeneous interests and intrinsic worth of every single individual. With respect to human and non-human living beings the concept of diversity is successful, because it conveys respect and responsibility, tolerance and pleasure of heterogeneity. Since the 1980s, 'diversity' has become a central concept in social emancipation movements. It emphasizes cultural difference and includes a critical reflection of one's own cultural-relative standpoint. But, again, 'diversity' functions by integration because it refers not only to current concerns but has also a deeply rooted historical dimension leading far back to the very first written texts of mankind (see Sect. 16.3). The story of biodiversity has been so successful because it is related to some deeply rooted ideals about the world, not least the idea of paradise: for one essential characteristic of the Garden of Eden is that it is full of plants and animals of different species coexisting in a joyful and peaceful manner.

Its ongoing scientific usage and at the same time latent connection to ancient cultural-religious ideas (such as paradise) makes 'biodiversity' a powerful concept for mediating between science and the broader public. It is a paradigmatic example of what has been called "post-normal science", where "facts are uncertain, values in dispute, stakes high and decisions urgent" (Funtowicz and Ravetz 1992, p. 138). This characterization applies particularly well to the status of 'biodiversity': First, the investigation of biodiversity has to cope with uncertainties on the factual as well as the axiological or ethical level. We simply do not know enough about the amount and function of biological diversity; we do not know, for example, whether there are currently three or 100 million species of animals on earth, and we do not know how they contribute to the stability of our ecosystems. Second, we do not know how we should evaluate biodiversity: instrumentally or intrinsically. Third, stakes are high because we are currently facing a loss of biological species probably on the level of one of the five mass extinction events in earth history. Finally, decisions are urgent because this is an irreversible loss and we do not know whether there will be a tipping point when things get worse at an increased speed and scale.

Furthermore, biodiversity studies are paradigmatic for post-normal science because they examine a field where laypersons are becoming experts. Big data provided by millions of people taking part in observation surveys particularly for birds and insects is an important basis for decision-making in conservation biology. These extended peer communities with their socially distributed expertise are especially important for the knowledge of local conditions.

These factors have transformed biodiversity studies from the old paradigm of scientific discovery ('Mode 1'), characterized by the hegemony of theoretical and experimental science, to a new paradigm of knowledge production (the 'Mode 2'), in which knowledge is "socially distributed, application-oriented, trans-disciplinary, and subject to multiple accountabilities" (Nowotny et al. 2003, p. 179). The investigation of biodiversity is a post-normal Mode 2-science because it is "issue-driven" and "mission-oriented" rather than theoretical and driven by curiosity. In situations of Mode 2-science experts are incapable of providing conclusive answers to the associated problems. They can provide their views but decisions have to be made in public forums such as the Intergovernmental Science-Policy Platform on Biodiversity and Ecosystem Services (IPBES).

Being immersed in issues of philosophy, cultural history and economy, biodiversity is not being a merely scientific problem anymore, but rather one where science and politics meet. As an object of public attention and a focal point of conflicting interests, 'biodiversity' concerns the management of an issue rather than the solution to a problem.

16.2   On Defining 'Biodiversity'

As the concept of 'biodiversity' is vague or versatile, it still has to be defined to be of any value in public or scientific argumentation. In the mid-1980s 'biodiversity' was explicitly introduced with a non-scientific, but political intent. The term was coined in preparation of the *National Forum on BioDiversity,* which took place in September 1986 in Washington, D.C. The American botanist Walter G. Rosen who was involved in preparing the conference explained how he came up with the term in a later interview: Creating the term, he said, was "easy to do: all you do is to take the 'logical' out of 'biological'" (Rosen 1992, in Takacs 1996, p. 37). Linguistically speaking it was an easy task: Rosen simply used 'biodiversity' instead of 'biological diversity' which was already a well-established technical term in biology. His aim was, as he said, to create room for "emotion" and "spirit".

And this is the situation we are in now: 'Biodiversity' is a term full of emotion and spirit, expressing an ethical concern following the mass-extinction of species due to human actions. Compared to this strong ethical pulse, the explicit definitions of the term that have been given after the conference in Washington are rather weak. Most of them given by biologists, for example, are very broad. In one of the first explicit definitions of 'biodiversity' formulated by the US Office of Technology Assessment (OTA) in 1987, it is defined as "the variety and variability among living organisms and the ecological complexes in which they occur" (OTA 1987, p. 3). A few years later, Solbrig's much cited definition explains the term as "the property of living systems of being distinct, that is different, unlike" (Solbrig 1991, p. 9). The

Convention on Biological Diversity from 1992 sees it as "the variability among living organisms from all sources" (UNEP 1992, art. 2). In one of the implicit and open definitions Sarkar explains biodiversity as "what is being conserved by the practice of conservation biology" (Sarkar 2002, p. 132), and a few years later, Norton demands that any definition must be rich enough to capture "all that we mean by, and value in, nature" (Norton 2006, p. 57).

From these definitions it is obvious that the term was designed to be open, versatile, polyvalent and adaptable to changing situations (Casetta and Delord 2014, p. 251). It has been characterized as an "umbrella concept" encompassing the entire field that has formerly been called "nature protection" or "nature conservation" (Potthast 2014, p. 132). Because of its oscillation between scientific and non-scientific contexts Gutmann sees it as a "pragmatic concept" or "metaphor" (2014, p. 66). Regular movements across discursive borders allow the concept to touch upon and somehow integrate many diverse aspects of nature and its use by humans. This mediating quality depends on the term's "performativity" by mobilizing attitudes and reactions in diverse contexts (Casetta and Delord 2014, p 251). For this performativity to be effective, the multifarious character of 'biodiversity' is essential; it allows for the fact that "[i]n biodiversity each of us finds a mirror for our most treasured natural images, our most fervent environmental concerns." (Takacs 1996, 81).

Thus, 'biodiversity' is exactly what a politically successful concept ought to be: sufficiently open in order to be meaningful to many people and powerfully employed in political processes. It amalgamates scientific and political developments, public concern and cultural changes in society.

16.3   Representing Biodiversity

Apparently, the success of 'biodiversity' as a concept in the public discourse results, at least in part, from its reference to political and social concerns about diversity in non-biological contexts. Another reason for its power in social discourse might be that it brings into play ancient formats of representing the multitude of things. Representing biodiversity—understood as species' inventory—starts with the beginning of writing, roughly 5000 years ago. In the world's earliest examples, which are lists from Mesopotamian cultures, a huge variety of trees, domestic animals, fish and birds appear alongside lists of goods which have been traded, as well as other inventory-like lists of things in the world: metals, vessels, official functions, and geographic places (Veldhuis 2014). The early list of bird species in proto cuneiform from around 3000 BC featured more than 100 entries, including ravens, which presumably were of no direct use or benefit to humans. The lists seem to be inventories of species or of kinds of things in the world regardless of their utilitarian value.

Since Mesopotamian times, such lists have been used by natural historians to log, check and order species of living beings. Linnaeus' *Systema naturae* is basically still a list; in its tenth edition of 1758 it contains about 4200 species on 800 pages. Today's records of diversity are also organized as lists, for example the *Encyclopedia of Life,* the largest online database of systematic biology.

The discovery of biodiversity could be seen as a science of lists—*Listenwissenschaft* as Assyriologists have called the presentation of knowledge in this form in Mesopotomian cultures (Schneider 1907, p. 368). The list's essential logic is paratactical: lists do not primarily *explain,* as hypotactical theory-centered science does, but they first of all describe and arrange things on the same level. Lists are apt devices for the egalitarian, non-hierarchical presentation of things.

Biodiversity images put this logic into the visual sphere by showing an egalitarian representation of diverse living beings. Paradigmatic images of this type appear in the Flemish still life of the late sixteenth and seventeenth century, for example in the work of Joris Hoefnagel or Jan van Kessel the Elder. The scattered arrangement of decontextualized animal bodies, a "strewn pattern image" (Schütz 2002, p. 66), shows what biodiversity essentially is: a state of difference. Biodiversity is the sum of different individuals with different lifeforms and lifestyles. They are presented according to the principle of addition, a colorful diversity that does not manifest a closed totality or a system of interaction. Central to this depiction is that there is no top or bottom, no hierarchy and no evaluation within one set. The principle of representation is a paratactic egalitarianism, the line-up of individuals with an equal standing, a juxtaposition of forms.

In the popular visual culture of our days you can find many examples of images that follow this paratactic, egalitarian logic. One example is a photo project by nature photographer David Liittschwager: In *A World in One Cubic Foot* he took a bright green metal cube—measuring precisely one cubic foot—and set it in various ecosystems around the world, from Costa Rica to New York Central Park (Liittschwager 2012). He documented what moved through that small space in a period of 24 h and photographed the plant and animal life he encountered in that period of time. An image of local biodiversity was then created by compiling all these photographs according to paratactic logic. In another example, Christopher Marley arranged his photographs of beetles in a kind of biodiversity mosaics (Marley 2008). This resulted in holistic figures such as squares or circles. No mosaic stone here resembles the other, and no element may be missing for the whole to be complete.

There is a striking parallel between the iconic logic of contemporary biodiversity pictures and some seventeenth-century Flemish still lives. The still lives, however, do not visualize ethical concerns about extinction of species, but refer to their creation. Their reference point is natural theology: the animals are considered immediate manifestations of God's will and thus provide access to God's plan equal to the Holy Scriptures. Although the modern concern with extinction and the late seventeenth-century focus on creation are rather distinct, they have one essential thing in common: the emphasis on and evaluation of individuals and species. In biodiversity images it is individuals and species that are appreciated in the first place, whereas their interactions with their environment and each other are pushed into the background. This decontextualizing, egalitarian logic, "specimen logic" as Jenice Neri has called it (Neri 2011, p. xiii), is essential for understanding our conceptualization and appreciation of biodiversity.

In the last decades, this specimen logic, the paratactic order of individuals, has become a dominant mode of presenting animals. It is manifest in installations at natural history museums such as the *Hall of Biodiversity* in the American Museum of Natural History in New York or the *Wall of Biodiversity* in the Museum für Naturkunde in Berlin. Without further explanation, these installations group together a great variety of stuffed animals that do not naturally occur next to each other in the same location. The installations are not about explaining and understanding, but about creating astonishment and amazement by this variety. The focus is on the aesthetic quality of the individual objects and on the feeling that each species is threatened by humans. Thus, questions of nature conservation and ethics are addressed. In parallel with older biodiversity images, the museum installations are characterised by the fact that (1) they stress the individual character of different species by presenting them in an extremely naturalistic way, (2) they decontextualize each species by displaying only one individual devoid of its ecological setting, and (3) they arrange the specimens in a non-hierarchical, egalitarian order. In contrast to earlier forms of presentation in natural history museums, biodiversity installations abandon showing causal, functional or systemic relations. The rationale of this form of presentation is that the mere sequence of different animal bodies is intended to be free of ideological or cultural preconceptions.

In these museum installations, the individuals play the central role: Each specimen not only represents the living organism its parts once belonged to, but, as there is only one specimen for each species, each specimen also represents its entire species. In these representations biodiversity is an ethic not for individuals in the first place but for species; it is about the moral dignity of species. The installation exemplifies yet another form of representation, a political representation: If the showcase of the installations is seen as a kind of parliament then each species has one vote in it; there is an equal representation for each species.

Interpreted in this way, the representations of biodiversity in natural history museums correspond, of course, to the normative discourse of egalitarian pluralistic democracies. Accordingly, it may be seen in terms of political iconography: as an expression of pluralistic social and political ideas. In short, it displays political ideals in the guise of nature.

However, the non-hierarchical, paratactic logic of presenting animals not only corresponds to liberal ideas of an egalitarian society, it also corresponds to the store-aesthetics of the market economy where the consumer can choose among the many products of equal value offered to him as being different. The increased attention to diversity can thus also be interpreted as having been influenced or enforced by capitalist economy. Moreover, it has also been argued that the origin of the dominating "taste for colorful diversity" lies in "the market": "It is the taste formed by the contemporary market, and it is the taste for the market" (Groys 2008, p. 151).

The aesthetic of diversity thus has many different roots. It can be found, amongst others, in the very old history going back to the Mesopotamian *Listenwissenschaft*, the general pleasure in the manifold of Western culture (*poikilia* in Greek, *varietas* in Latin; Grand-Clément 2015; Fitzgerald 2016), the social emancipatory movements of the second half of the twentieth century, capitalist market economy, or just postmodern taste. Just as 'biodiversity' refers to a multitude of perspectives that cannot be reduced to one coherent system there is no master-narrative for the explanation of its current success.

16.4   The Hybridization of Facts and Values in 'Biodiversity'

The various new and old traditions of diversity have in common that they are not merely descriptive but place value on variety and variability. The intersection of fact and value in the representation of diversity is particularly evident with respect to biodiversity. In the Judeo-Christian context four fundamental scenes are connected to biodiversity: scenes of Creation, of Paradise, the naming of animals by Adam and the animals boarding Noah's ark. As these are well-known religious scenes in a particular ethical context, their effect is placing values on diversity, charging it positively. This evaluative stance is also evident in Christopher Columbus's first letter from the New World (addressed to the finance minister at the Spanish court, Luis de Santángel, who supported Columbus) in which reference to the biodiversity he encountered—trees and birds of "a thousand different kinds" (*de mil maneras*)—forms an essential element of his praise of the promising land he discovered.

A more explicit appreciation of diversity can be found in Christian authors such as Augustine of Hippo who used reference to the huge diversity of animal species ("tantas diversitates animalium") for the praise of God: "how great all these things are, how magnificent, how beautiful, how amazing! And he who made them all is your God" (*Enarrationes in Psalmos*, 145, 12; transl. by Boulding 2004, p. 411). In a similar vein, Thomas Aquinas wrote: "Although an angel, considered absolutely, is better than a stone, nevertheless two natures are better than one only; and therefore a universe containing angels and other things is better than one containing angels only; since the perfection of the universe is attained essentially in proportion to the diversity of natures in it, whereby the diverse grades of goodness are filled, and not in proportion to the multiplication of individuals of a single nature." (*Scriptum super Sententiis*, lib. I, dist. 44, quaest.1, art 2; transl. by Lovejoy 1936, p. 77).

Similar views are expressed in the writings of Leibniz, who weighs one man against the whole species of lions, and writes, by the way, that he is not sure whether God would actually prefer the individual human (*Essais de théodicée*, 1710; Lovejoy 1936, p. 225). In this understanding of diversity, human beings are co-ordinated with the other species of living beings; they do not inhabit an absolute and excellent position, but compete eye-to-eye with other species. Value is placed not (only) on the intrinsic qualities of each single being but on the state of being different from others. This evaluative stance towards diversity as such can be seen as a prefiguration of the modern concept of biodiversity as an "epistemic-moral hybrid" (Potthast 2014, p. 138); it is a kind of biodiversity *avant la lettre*.

On the basis of these prefigurations it was an easy task for concerned biologists in the late twentieth century to propagate 'biodiversity' as an important issue by taking out the 'logical' from 'biological diversity'—which was taken to be a biological term—, and putting in "emotion" and "spirit" instead. This strategic reenchantment of a (supposedly) biological concept made the term very useful for the political sphere. Being full of concern and sufficiently vague, open to many ideas, even contradictory ones, the term became an effective instrument for politics.

However, in recent years the intrinsic value and hybrid character of 'biodiversity' came under attack. It has been doubted that biodiversity is a useful basis for decisions in nature conservation (Morar et al. 2015; Santana in this book). For in many cases we are not interested in the diversity of things as such. Nature conservation is often concerned not with protecting as many species as possible, but only very specific, typical or rare ones. In some cases, we are trying to limit genetic diversity, for example in cases where it leads to functional disorders, or we are trying to eradicate pathogens. Diversity in itself is not always a good thing, but only the right measure of it, so it has been argued.

Accordingly, one problem of the concept is the unconditionally positive evaluation of diversity. Another problem is that the evaluative charge of the term 'biodiversity' makes it impossible to distinguish between scientific knowledge as such and the process of evaluating this knowledge. Morar et al. (2015) argue that we should decide in an open democratic discourse which diversity we want where. The amalgamation of the two steps of gaining and evaluating knowledge into one, as the hybrid concept 'biodiversity' does, obscures the need for separating scientific facts and public review of its results: "the role of [political] judgments is obscured when decisions are presented as following automatically from empirical evidence" (Morar et al. 2015, p. 25).

This criticism problematizes exactly that aspect of the concept, which was responsible for its success: the hybridization of descriptive and normative dimensions. Obviously, the comprehensive success of the rhetoric of biodiversity was bought at no small price. Its power to hybridize makes biodiversity a useful political concept but it also stands in the way of any precise argument. Good intentions and positive effects connected to the concept cannot replace differentiated ethical reasoning. The important integrative function of the concept needs to be complemented with arguments in which the hybridization of facts and norms, of science and values, of knowledge and wonder is carefully separated again. Not scientists and their concepts but the democratic society as a whole has to decide which diversity it desires, one that includes *genetic disorders*, the *polio virus* or the *Anopheles mosquito* – or not.

16.5   Conclusion: Biodiversity as an Absolute Metaphor

Because of its involvement in various ancient traditions, and correspondingly hybrid and multifarious character, 'biodiversity' can be understood as an "absolute metaphor" or "non-concept" (*Unbegriff*) in the sense of Hans Blumenberg. Just as 'life', 'time' or 'world' the word cannot have a fixed meaning because it mediates between various contexts and disciplines. Aspects of the term can be defined within each separate context. Biology, for example, has provided many mathematically precise definitions of biological diversity as an index for measuring species richness and evenness in their distribution (but, already at this level diversity has been called a "non-concept" because it can be measured in very different ways; see Hurlbert 1971).

However, this technical understanding of biological diversity is distinct from 'biodiversity' as it functions in public debates. In these debates the concept functions as an absolute metaphor the meaning of which cannot be exhausted in any context of its use and proves "resistant to terminological claims and cannot be dissolved into conceptuality" (Blumenberg 2010, p. 5). It therefore seems that any attempt to define 'biodiversity' in precise terms is futile and does not capture the role that the word fulfills in contemporary discourses. Its interdiscoursive function depends on not having a clear-cut definition but being an open concept. 'Biodiversity' not only refers to the variety and variability of the natural world but also to our conceptualization and valuation of it. This complexity is the reason for the vagueness and at the same time the power of the concept: "le plus vague est le plus puissant" (Bachelard 1947, p. 184).

References

Augustine of Hippo. (2004). *Enarrationes in Psalmos* (Vol. 6, Trans. and notes by M. Boulding). Hyde Park: New City Press.

Bachelard, G. (1947). *La formation de l'esprit scientifique*. Paris: Vrin.

Blumenberg, H. (2010). *Paradigms for a metaphorology (1960)*, (R. Savage, Trans.). Ithaca: Cornell University Press.

Casetta, E., & Delord, J. (2014). Versatile biodiversité. In E. Casetta & J. Delord (Eds.), *La biodiversité en question. Enjeux philosophiques et scientifiques* (pp. 247–253). Paris: Éditions Matériologiques.

Díaz, S., et al. (2015). The IPBES conceptual framework – Connecting nature and people. *Current Opinions in Evironmental Sustainability, 14*, 1–16.

Fitzgerald, W. (2016). *Variety. The life of a Roman concept*. Chicago: University of Chicago Press.

Funtowicz, S. O., & Ravetz, J. R. (1992). A new scientific methodology for global environmental issues. In R. Costanza (Ed.), *Ecological economics. The science and Management of Sustainability* (pp. 137–152). New York: Columbia University Press.

Grand-Clément, A. (2015). Poikilia. In P. Destrée & P. Murray (Eds.), *A companion to ancient aesthetics* (pp. 406–422). Chichester: Wiley Blackwell.

Groys, B. (2008). *Art power*. Cambridge, MA: MIT Press.

Gutmann, M. (2014). Biodiversity: A methodological reconstruction of some fundamental misperception. In D. Lanzerath & M. Friele (Eds.), *Concepts and values in biodiversity* (pp. 55–72). London: Routledge.

Heyd, D. (2010). Cultural diversity and biodiversity: A tempting analogy. *Critical Review of International Social and Political Philosophy, 13*, 159–179.

Hurlbert, S. H. (1971). The nonconcept of species diversity: A critique and alternative parameters. *Ecology, 52*, 577–586.

Liittschwager, D. (2012). *A world in one cubic foot. Portraits of biodiversity*. Chicago: University of Chicago Press.

Lovejoy, A. O. (1936). *The great chain of being*. Cambridge, MA: Harvard University Press.

Marley, C. (2008). *Pheromone*. Portland: Pomegranate Communications.

Millennium Ecosystem Assessment. (2005). *Ecosystems and human well-being: Synthesis*. Washington, DC: Island Press.

Morar, N., Toadvine, T., & Bohannan, B. J. M. (2015). Biodiversity at twenty-five years: Revolution or red herring? *Ethics, Policy & Environment, 18*, 16–29.

Neri, J. (2011). *The insect and the image. Visualizing nature in early modern Europe, 1500–1700*. Minneapolis: University of Minnesota Press.

Norton, B. (2006). Toward a policy-relevant definition of biodiversity. In J. M. Scott, D. D. Goble, & F. W. Davis (Eds.), *The endangered species act at thirty* (Vol. 2, pp. 49–58). Washington, DC: Island Press.

Nowotny, H., et al. (2003). Mode 2 revisited: The new production of knowledge. *Minerva, 41*, 179–194.

Office of Technology Assessment (OTA). (1987). *Technologies to maintain biological diversity*. Congress of the United States, OTA-F-330.

Pope Francis. (2015). *Laudato si'. Encyclical letter of the holy father Francis on care for our common home*. http://w2.vatican.va/content/francesco/en/encyclicals/documents/papa-francesco_20150524_enciclica-laudato-si.html

Potthast, T. (2014). The values of biodiversity. In D. Lanzerath & M. Friele (Eds.), *Concepts and values in biodiversity* (pp. 132–146). London: Routledge.

Sarkar, S. (2002). Defining "biodiversity", assessing biodiversity. *The Monist, 85*, 131–155.

Schneider, H. (1907). *Kultur und Denken der alten Ägypter* (Vol. 1). Leipzig: Voigtländer.

Schütz, K. (2002). Naturstudien und Kunstkammerstücke. In C. Nitze-Ertz (Ed.), *Sinn und Sinnlichkeit. Das flämische Stillleben (1550–1680). Eine Ausstellung der Kulturstiftung Ruhr Essen und des Kunsthistorischen Museums Wien* (pp. 60–109). Lingen: Luca-Verlag.

Solbrig, O. T. (1991). *Biodiversity. Scientific issues and collaborative research proposals*. Paris: UNESCO.

Takacs, D. (1996). *The idea of biodiversity. Philosophies of paradise*. Baltimore: Johns Hopkins University Press.

UNEP. (1992). *Convention on biological diversity*. https://www.cbd.int/convention/

UNESCO. (2001). *Declaration on cultural diversity*. http://portal.unesco.org/en/ev.php-URL_ID=13179&URL_DO=DO_TOPIC&URL_SECTION=201.html

Veldhuis, N. (2014). *History of the cuneiform lexical tradition*. Münster: Ugarit.

17

Defining Biodiversity and the Lack of Transparency

Yves Meinard, Sylvain Coq, and Bernhard Schmid

Abstract The vagueness of the notion of biodiversity is discussed in the philosophical literature but most ecologists admit that it is unproblematic in practice. We analyze a series of case studies to argue that this denial of the importance of clarifying the definition of biodiversity has worrying implications in practice, at three levels: it can impair the coordination of conservation actions, hide the need to improve management knowledge and cover up incompatibilities between disciplinary assumptions. This is because the formal agreement on the term "biodiversity" can hide profound disagreements on the nature of conservation issues. We then explore avenues to unlock this situation, using the literature in decision analysis. Decision analysts claim that decision-makers requesting decision-support often do not precisely know for what problem they request support. Clarifying a better formulation, eliminating vagueness, is therefore a critical step for decision analysis. We explain how this logic can be implemented in our case studies and similar situations, where various interacting actors face complex, multifaceted problems that they have to solve collectively. To sum up, although "biodiversity" has long been considered a flagship to galvanize conservation action, the vagueness of the term actually complicates this perennial task of conservation practitioners. As conservation scientists, we have a duty to stop promoting a term whose vagueness impairs conservation practice. This approach allows introducing a dynamic definition of "biodiversity practices", designed to play the integrating role that the term "biodiversity" cannot achieve, due to the ambiguity of its general definition.

Y. Meinard (✉)
Université Paris-Dauphine, PSL Research University, CNRS, UMR [7243], LAMSADE,
Paris, France
e-mail: yves.meinard@lamsade.dauphine.fr

S. Coq
Centre d'Écologie Fonctionnelle et Évolutive (CEFE), CNRS, Montpellier, France
e-mail: Sylvain.COQ@cefe.cnrs.fr

B. Schmid
Institute of Evolutionary Biology and Environmental Studies, University of Zurich,
Zurich, Switzerland
e-mail: bernhard.schmid@ieu.uzh.ch

Keywords Biodiversity · Definition · Conservation practice · Problem solving ·
Decision analysis

17.1   Introduction

The Convention on Biological Diversity defines biological diversity or biodiversity
as "the variability among living organisms from all sources including, inter alia,
terrestrial, marine and other aquatic ecosystems and the ecological complexes of
which they are part: this includes diversity within species, between species and of
ecosystems (United Nations 2013)." This now classical definition is largely discussed in the philosophical literature for being exceedingly vague and in need of
clarification (Sarkar 2005; MacLaurin and Sterelny 2008; Meinard et al. 2014;
Santana 2014). By contrast, most ecologists consider that this vagueness is unproblematic in practice (Mace et al. 2012).

   In this chapter, we argue that this vagueness does have worrying implications in
practice, at three levels: it can impair the coordination of conservation actions, hide
the need to improve management knowledge and cover up incompatibilities between
disciplinary assumptions. Our purview in this chapter is accordingly mainly practical: we aim to address ecologists, conservation biologists and practitioners, with the
objective of convincing them that debates on the definition of biodiversity may have
concrete implications. The problems that we thereby highlight all stem from the
lack of a clear and shared definition of biodiversity. Biodiversity is certainly not the
only concept that suffers from being vaguely defined, and in many cases this vagueness does not create much problems. Accordingly, our aim here is not to claim that
vagueness is a problem in itself. Our more modest aim is to argue that, in the very
specific case of biodiversity, it does have worrying consequences.

   Indeed, in this specific case, formal agreements among various actors on the term
"biodiversity" can hide profound disagreements on the nature of conservation and
ecological issues. This is reminiscent of a classical problem in decision modelling
(Bouyssou et al. 2000), for which the proven solution is for interacting actors to
articulate a commonly accepted formulation of the key questions structuring their
interaction. In line with this view, we propose that, although the notion of biodiversity does not unify biodiversity sciences in a transparent, rigorous way, such a unification may be achieved by clarifying a concept of *biodiversity practices*, understood
as coherent collaborative interdisciplinary efforts to tackle commonly identified
environmental and conservation problems. We take advantage of insights from the
philosophical literature to champion this approach and to argue that, although a
definitive definition of these biodiversity practices might be unreachable, the task to
constantly improve definitions, taking seriously conservation biologists' and conservation practitioners' value-laden stances, is crucial to the enrichment and
improvement of conservation theories and practices. If we may paraphrase Burch-
Brown and Archer (2017), although we emphasize that one cannot hope to reach a
definitive answer to the question "what is biodiversity?", our approach hence proposes a "defense of biodiversity" that consists in championing a collective effort to

constantly improve our understanding of the value-laden practices gathered under banner of "conserving biodiversity".

The remainder of this chapter is divided into three parts. In Sect. 17.2, we first show that, despite its being seemingly simple and unequivocal, the definition of "biodiversity" is exceedingly vague. Vagueness in itself is not necessarily a problem. But Sect. 17.3 uses cases studies to show that, in the case of "biodiversity", this vagueness creates problems in practice. In Sect. 17.4 we then explain our proposed solution. Section 17.5 briefly concludes.

17.2   The False Transparency of the Definition of Biodiversity

The vagueness of definitions of "biodiversity" has been extensively studied in the philosophical literature (Sarkar 2005; MacLaurin and Sterelny 2008; Meinard et al. 2014), but for lack of a concrete understanding of its implications for conservation science and practices, this debate has been largely confined to philosophical discussions without affecting real-life conservation and ecological practices (for a noticeable exception, see Delong 1996). Let us first explain why we claim that definitions of "biodiversity" are vague.

17.2.1   Diverging Definitions of "Biodiversity" Coexist

A first example will illustrate how deceptive is the idea that the definition of "biodiversity" is clear and unequivocal. Let us look at two prominent approaches to biodiversity, articulated by a leading author in conservation biology and a leading author in ecosystem ecology: Sarkar (2005) and Loreau (2010).

Loreau (2010) does not delve into definitional debates. He uses a definition very similar to the one of the CBD, stating that "biodiversity [...] includes all aspects of the diversity of life—including molecules, genes, behaviors, functions, species, interactions, and ecosystems" (p. 56). The fact that he uses such a sketchy definition suggests that he takes the definition to be unproblematic and consensual. By contrast, Sarkar (2005) explicitly tackles the definitional issue. Following Maclaurin and Sterelny (2008, p. 8), one can summarize his approach by stating that, according to his definition, "'biodiversity' [means] whatever we think is valuable about a biological system" (Maclaurin and Sterelny's interpretation of Sarkar's theory can be criticized, but for the purpose of the present chapter, we will not delve into this exegetic debate).

A striking difference between Loreau's (2010) and Sarkar's (2005) definitions is that, whereas Sarkar's definition explicitly mentions values, Loreau's definition exclusively mentions purely biological concepts and objects. Despite this major difference, Sarkar explicitly claims that he use the concept of biodiversity in an uncontroversial and widely shared sense: he even writes that his approach captures the "consensus view" (p. 145). And Loreau makes the same claim, though implicitly, since he admits that there is no need to delve into definitional issues. Despite the major difference between their respective definitions, both authors hence claim that their approach captures the general understanding of the concept.

Hence, although Loreau (2010) and Sarkar (2005) use the same term and take for granted that they understand it in the same way as everyone else, they actually understand it markedly differently. Can this kind of misunderstanding have practical implications? In the sections to follow, we argue that, in the case of biodiversity, it can.

17.2.2 The Various Disciplinary Studies "of Biodiversity" Do Not Study the Same Things

The literature presenting the numerous measures and indexes of biodiversity is extensive (Muguran and McGill 2011). It is commonplace to notice that the different disciplines (encompassing what will thereafter be termed various "biodiversity studies") respectively favor different indexes because they capture concepts that are better adapted to their subject-matter. The term "biodiversity" is used in articles from these various disciplines mostly in introductions and conclusions, whereas discipline-specific concepts such as species richness (Fleishman et al. 2006), phylogenetic distances (Faith 1992) or functional traits or attributes (Petchey and Gaston 2002; Mason et al. 2003) replace it in the methods and results sections (Meinard 2011). Similarly, environmental economists often use the term "biodiversity" to introduce and justify their research, but rapidly switch to disciplinary concepts, such as "naturalness" (Eichner and Tschirhart 2007) or "perceived diversity" (Moran 1994). The same is true of the other disciplines concerned with biodiversity. Accordingly, although they all claim to study biodiversity, the various biodiversity studies actually produce results that account for different objects, properties and processes (Maclaurin and Sterelny 2008).

The concept of biodiversity itself is never used in articulating results, in any of these disciplines. It is mostly confined to introductions and conclusions, where it plays the role of a catchword.

17.2.3 The Various Disciplinary Studies "of Biodiversity" Presuppose that they Study Various Aspects of a Common Entity

By using the notion of biodiversity in their introductions and conclusions, all these heterogeneous studies presuppose, at least implicitly, that the various objects, properties and processes that they study are aspects of a common entity: biodiversity (here we use the term "entity" in a purportedly very large sense, encompassing all sorts of ontological units, such as objects, properties, natural kinds, and so on). They do not claim that their concepts or measures represent all of biodiversity, but that there is a common entity, biodiversity, which is partially captured by their favorite measures and concepts.

In the current literature on biodiversity, the various studies simply state that their subject-matter is an aspect of the putative entity biodiversity, without explaining what this entity is supposed to be. What is this putative common entity supposed to be?

17.2.4 Defining "Biodiversity" Thanks to the Notions of Diversity or Variety Is Insufficient to Identify such a Common Entity

The literature on indexes and measures of biodiversity is notably vague on the issue of a proper identification of this common putative entity—biodiversity. The usual explanation identifies it as a specification of a more general entity: the property diversity (Maris 2010). Biodiversity would be the diversity of living things (Gaston and Spicer 2004), along genetic, phylogenetic and functional dimensions (Purvis and Hector 2000).

This approach bears some seeming credibility because "diversity", and synonyms in ordinary language such as "variety", belong to the everyday language and thus seem clear and self-evident. Intuitively, diversity is a property characterizing groups of individuals, depending on the number of individuals and on their similarities and dissimilarities. But the precise roles of numbers, similarities and dissimilarities, and the metrics used to measure them, are not elucidated at this intuitive level.

To determine whether "diversity" truly captures a coherent notion, axiomatic studies have tried to formalize the properties associated with it (Weitzman 1992; Nehring and Puppe 2002). They thereby showed that these properties are highly variable and that the notion of diversity is accordingly deeply ambiguous (Gravel 2008) (in other words, what these studies show is that, whereas it seems self-evident at first sight that diversity is a property, in fact the term "diversity" captures different sets of properties in different contexts, which makes it questionable to claim that "diversity" refers to a property properly speaking). The terms "diversity" or "variety" thus function like a term such as "adaptation". "Adaptation" has different meanings in various subfields of evolutionary biology, it has a markedly different meaning in medical physiology, and yet other meanings in ordinary language. The same holds true for "diversity" and "variety". Within disciplines or, more precisely, within subfields, these terms are relatively unambiguous and generally well-defined, but their meaning varies between disciplines or subfields. As a consequence, these terms cannot be unambiguously used in both ways at the same time. Either one relies on subfield-specific, technical and well-clarified definitions of the terms "diversity", "variety", etc.—but in that case one can no longer draw upon the self-evidence of these terms in everyday language. Or one relies on everyday language—but in that case, one has to face the fact that ordinary language does not delineate coherent notions of diversity or variety. In both cases, using the terms "diversity" or "variety" in a general definition of biodiversity is problematic, because one cannot take for granted that others will understand the notion in the intended way. Therefore, if buttressed on general terms like "diversity" or "variety", a general definition of biodiversity does not single out a unique entity, and is therefore useless to support the idea that "biodiversity" refers to a common entity.

Here again, the comparison with "adaptation" is illustrative. A rigorous evolutionary biologist would never use the term "adaptation" when talking to lay people or to physiologists without specifying that his technical understanding of the term is very specific. The evolutionary biologist knows that his interlocutors think that they understand the term "adaptation", and he knows also that, in a sense, they are right to think that they understand the term. But he also knows that they understand the term in another sense, rather than the one he has in mind. Therefore, it is natural for him to clarify the meaning of the term. This crucial step is the one that is missing in the case of "biodiversity".

The theoretical considerations developed in this Sect. 17.2 may appear purely formal, without implications for concrete conservation science and action. The goal of the following section is to demonstrate that the reverse is true.

17.3   How False Transparency Creates Concrete Problems for Conservation Science and Action

In order to explain the concrete problems created by the seemingly purely theoretical reasoning spelled out in Sect. 17.2, let us now take three concrete case studies, each illustrating a specific kind of problem.

17.3.1   The False Transparency of "Biodiversity" Can Impair the Coordination of Interacting Conservation Actions

Misunderstandings created by the false transparency of "biodiversity" can have detrimental implications at the level of practical conservation management, as can be illustrated by the story of the management of the Bel-Air valley in South-west France (Gereco, unpublished report 2014). This is a small valley (Fig. 17.1) containing a rich mosaic of aquatic and humid habitats in a karstic system close to semi-arid grasslands and upstream water meadows (surrounding the Charente River).

This valley shelters a population of otters (*Lutra lutra*) and a massive population of Louisiana crayfish (*Procambarus clarkia*). The latter is an invasive species having major detrimental impacts on the functioning of aquatic ecosystems (Angeler

Fig. 17.1 The Bel-Air valley

et al. 2001; Rodriguez et al. 2005). However, its impact on Mammals populations is modestly positive (Correira 2001), and from the point of view of otter-watchers it has the advantage to turn otters' spraints into red, greatly facilitating the observation and monitoring of otter populations. The above report also unveiled the presence of Japanese knotweed (*Reynoutria japonica* Houtt., 1777), an invasive plant with deeply damaging impacts on wetland ecosystems.

The valley is managed by an environmental association, Perennis. The downstream water meadows are protected under the Habitat Directive (HD, a cornerstone of the European Union policy to maintain biodiversity: European Commission 1992) and are accordingly managed by another environmental association, the Birds Protection League ("LPO"). Both actors act according to management schemes explicitly aimed at conserving "biodiversity".

But on closer examination, it appears that Perennis understands "biodiversity" in a Sarkar-like manner. Indeed, as amateur naturalists, they value first and foremost the emblematic otters: for them promoting biodiversity mainly means managing the otter population. Because crayfish makes it easier to observe otter, and because they have not witnessed any impact of knotweed on otters yet, they do not see invasive species as a prominent topic in their agenda to conserve "biodiversity". By contrast, directed as it is by European guidelines applicable to the entire Natura 2000 network, the LPO has to conceive of its objective to preserve biodiversity in a way that puts more emphasis on ecological functioning. In particular, following the guidelines spelled out in Evans and Arvella (2011), its management actions have to actively tackle the problems created by invasive species populations. Accordingly, for the LPO, conserving biodiversity in this area implies managing the crayfish and knotweed populations (or at least it implies a need to carve out a strategy assessing the kind of invasive mitigation actions that can be performed, and the cogency of implementing them in the light of their cost and likelihood of success).

Perennis' management strategy aims at "conserving biodiversity", but this means protecting the otter population, and does not mean tackling the invasive species issue; similarly, the LPO's strategy aims at "conserving biodiversity", but this time it means tackling the invasive species issue. Both actors could agree when comparing their objectives: they both strive to "conserve biodiversity." But if it dismisses the invasive issue when managing the valley, Perennis actually jeopardizes any attempt to tackle this very issue downstream. The formal agreement on "biodiversity" hence hides a deep disagreement on what has concretely to be done.

At this stage, one might retort that misunderstandings like the one sketched above can easily be solved if the actors talk to each other about the concrete actions they want to implement. This is certainly true, and this example is indeed somewhat schematic. Our personal experience however suggests that, in real-life management situations, such seemingly trivial disagreements can persist. This is because the term "biodiversity" provides a common vocabulary that various actors can use to express very different objectives, which can all too easily lead them to fail to see the underlying divergences. In the present work, we obviously do not claim to have quantitatively demonstrated that such problems often arise in concrete conservation situation. Our more modest claim is that it can arise.

17.3.2   The False Transparency of "Biodiversity" Hides the Need to Improve Scientific Knowledge to Solve Complex Management Problems

The case of the Bel-Air valley provided a first illustration of how a concrete management problem can remain unseen because various actors fail to see the need to compare their respective understandings of "biodiversity". In this case, the problem arises at the level of the interactions between actors implementing conservation actions. But a deeper problem can arise when innovative solutions and new management knowledge are needed to solve more complex conservation issues. In such cases, the false transparency of "biodiversity" can hide the need to improve scientific knowledge.

An example illustrating this idea is given by the management of so-called "habitats of community interest", when biological invasion mitigation conflicts with habitat conservation (see Jeanmougin et al. 2016 for a deeper investigation of this conflict). "Habitats of community interest" (HCI) are natural or semi-natural habitats constituting the Natura 2000 network, as application of HD (European Commission 1992). HCI are typically defined in European guidelines (European Commission 2013) and more detailed regional scale manuals (e.g. Bensettiti 2001–2005) by lists of floristic species. For some HCI, these lists contain numerous invasive species (see Jeanmougin et al. 2016, SI-Table 3). For example, this is the case of the HCI "Constantly flowing Mediterranean rivers with *Paspalo-Agrostidion* species and hanging curtains of *Salix* and *Populus alba* (Habitat 3280)", whose presence has been recently reported in the lower Taravo River area (Corsica, France) (Fig. 17.2) (Gereco, unpublished report 2015).

Eight of the 34 index species of this habitat (*Paspalum distichum* L., 1759, *Paspalum dilatatum* Poir., 1804, *Xanthium strumarium* L., 1753, *Symphyotrichum subulatum var. squamatum* (Spreng.) S.D.Sundb., 2004, *Dysphania ambrosioides* (L.) Mosyakin & Clemants, 2002, *Amaranthus retroflexus* L., 1753, *Cyperus eragrostis* Lam., 1791 and *Erigeron canadensis* L., 1753) are considered invasive species according to various European, national or local databases. HD, as a politi-

Fig. 17.2 Paspalo-Agrostidion and curtain of Salix purpurea along the Taravo river

cal tool to maintain biodiversity, promotes the maintenance of HCI. On the other hand, the control and eradication of invasive species is also a central objective of many European initiatives to maintain biodiversity, such as the DAISIE (Delivering Alien Invasive Species Inventories in Europe) program and the recent European Directive on Invasive Species (Beninde et al. 2015).

In the case of habitats like HCI 3280, there is an antagonism between the invasive approach and the habitat approach. Indeed, if management actions achieve to mitigate populations of the above-cited invasive species, this will unavoidable imply that the area identifiable as HCI 3280 will decrease. Conversely, if management actions achieve an increase of the area occupied by HCI 3280, this will be accompanied by an expansion of populations of the above-cited invasive species. Consequently, elaborating a management scheme in areas like the lower Taravo is problematic, because two actions that are typically considered keystones of any biodiversity conservation strategy (invasive species mitigation and habitat conservation) are antagonist in such cases.

However, there is no scientific study or publication tackling this question (see Jeanmougin et al. 2016 for a bibliographic exploration quantitatively corroborating this idea). According to the database (ETC-BD 2015) constructed as part of the European-wide evaluation of the conservation status of HCI (European Union 2015), this habitat is present in no less than five countries in Europe (France, Greece, Italy, Spain and Portugal). Management schemes are hence devised and implemented all year long in the whole European Mediterranean region to manage this HCI, but there is no scientific guideline to decide how to solve the contradiction between the objectives to mitigate biological invasion and to promote the conservation status of habitat 3280.

Like most complex problems at the science-policy interface, this specific problem certainly has multifarious origins, having to do with the complex challenges in (1) translating ecological theory into practice (Knight et al. 2008), (2) defining the relevant expertise (Burgman et al. 2011), (3) choosing the relevant scientific paradigms to ensure operationality (Jeanmougin et al. 2016), (4) drawing the line between scientific information and advocacy (Brussard and Tull 2007), (5) assessing the proper place of scientific knowledge in the process of policy making (Josanoff 2012) and (6) entrenching the importance of an open diffusion of information on conservation practices (Meinard 2017a). We do not claim here to do justice to all these aspects, their interrelations and their relative importance in the genesis of problems such as the one of the above introduced lack of knowledge on HCI 3280 management and invasive species. Our more modest purpose is the following. We want to show that, by granting a key-role in the coordination between disciplinary approaches to a vague term like "biodiversity", one tends in all likelihood to render invisible the kind of knowledge gap at issue in our example. We accordingly do not claim to unfold a scientific demonstration here, but rather to hypothesize a possible mechanism that occurred to us thanks to our own field experience.

We propose that this mechanism is simply that specialists of invasive species stress the importance of controlling invasive species and present such a control as a prominent means to maintain biodiversity. But as non-specialists of habitats, they simply accept what specialists say about the self-evidence that maintaining HCI is also unquestionably good for biodiversity. Specialists of habitats behave in a symmetric way. Everyone thus agrees with the overarching objective to maintain biodiversity, everyone is careful not to question the expertise of one's interlocutor, and no one sees the need to improve knowledge and to carve out innovative management solutions in complex cases such as the one of habitat 3280.

As a consequence in the field, at the end of the story the resulting management scheme is most of the time decided more or less arbitrarily by political decision-makers or consultants on the basis of political, economic or circumstantial considerations. In the case of the Taravo River, the management scheme produced in 2014 (Lindenia, unpublished report 2015) does not mention this problem.

17.3.3   The False Transparency of "Biodiversity" Hides that Various Approaches Are Based on Incompatible Postulates

A more subtle, but no less important problem arises when interactions with non-ecological disciplines are involved. Let us start by illustrating the problem with an example: Eichner and Tschirhart (2007)'s ecological-economic study of a fishery ecosystem. Their aim is to establish how to organize fisheries given that the exploitation of a given species can have complex repercussions on the broader ecosystem. In their study system, human consumers buy items of one species (Pollock, *Theragra chalcogramma*) on markets and thereby indirectly impact other species due to between-species interactions in the ecosystem. This indirect impact then alters the provision of various ecological services. For example, the sheltering function of kelp (*Laminaria* spp.) can be altered, which has an impact on the populations of Pacific halibut (*Hippoglossus stenolepis*) and Pink salmon (*Oncorhynchus gorbusha*), which in turn alters the so-called "consumption services" (MEA 2005) for consumers of the latter species.

Eichner and Tschirhart (2007)'s solution is that if taxes on harvesting activities or caps on harvest, calibrated thanks to a precise knowledge of the functioning of the ecosystem, are implemented, demand will drop, overfishing will cease, kelp will recover, etc., and consumers will end-up being more satisfied.

The economists who authored this study claim that they provide insights that are complementary to those provided by ecologists to resolve a commonly identified problem—the problem of how to manage a complex ecological-economic system. Unfortunately, the way they see this problem is strikingly different from the way many ecologists see it. The compatibility of their prescriptions with prescriptions stemming from biological studies can accordingly become problematic.

Indeed, Eichner and Tschirhart (2007)'s understanding of the problem is based on a moral assumption—that is, an assumption about what is morally legitimate for scientists to do. They assume that consumers' preferences are given, and that the results of their study should not lead to a modification of consumers' preferences. Consequently, they do not integrate in their models the fact that being aware of the ecological impact of their act may change the behavior of the consumers. In technical economic terms, this assumption is encapsulated by the fact that human behavior is modeled using a predetermined utility function (Orléan 2011), whose parameters are not fed-back by the results of the study. Consumers are assumed to behave has if they were maximizing a function whose arguments are prices and quantities of goods they buy on markets. The knowledge of the system does not appear in the function: when a given consumer learns to know that his buying Pollock has impacts on populations of Salmon, this does not make any difference in his behavior on the Pollock market.

Such predetermined utility functions are often presented, like many other economic modelling tools, as morally neutral tools providing empirical complements to moral discussion (e.g. Scharks and Masuda 2016). When presented like this, it seems that predetermined utility function, as well as other modelling tools widely used in ecological-economic studies, can be used in conservation initiatives without interfering with the ethical motivations underlying the latter. Following the same logic, when an ecosystem ecologist works on a specific ecosystem process and an economist computes the economic value of an ecosystem service based on this process, it might look as though the two can work together and the end-result of their conjoint work is no less ethically neutral than the original ecological study of the ecological process. This repeatedly rehearsed logic is, however, largely acknowledged to be flawed: predetermined utility functions are not morally neutral modeling tools (Sen 2002; Hausman and McPherson 2006; Meinard et al. 2016). Using these tools means assuming that the results of the study should not lead to a modification of consumers' preferences: if consumers prefer x to y, the study should never aim at modifying this fact. This is not a technical constraint: implementing a feedback between the results of the model and preferences is not technically difficult (Lesourne et al. 2006). It is a moral stance: using predetermined utility functions is a means to promote the anti-paternalistic attitude to leave preferences as they are (Kolm 2005; Sagoff 2008) and to advocate that the satisfaction of preferences as they are is an acceptable, or even desirable, objective (Sagoff 2008).

This moral stance might seem reasonable enough—why should the economist think that he knows better than the consumer what the latter should prefer? But this moral stance can be problematic from the point of view of conservation sciences, because convincing people that their preferences are ill-conceived is a prominent means to achieve conservation targets. Many applied ecological studies are even openly based on moral assumptions that are diametrically opposed to the above one. Take for example the adaptive management approach (Norton 2005), according to which management practices should be seen as experiments from which managers can learn and thereby both improve their knowledge of the managed system and adjust the criteria that they use to evaluate alternative courses of actions (Lee 1993). Contrary to Eichner and Tschirhart (2007)'s model, this approach assumes that people's objectives and preferences are responsive to improvements of their knowledge of the ecological constraints (Maris and Bechet 2010), and that enabling such improvements is precisely one of the motivations to study these systems.

Eichner and Tschirhart (2007)'s solutions are solutions to the problem as they see it, constrained by moral assumptions that are not generally accepted by biologists. This does not mean that their approach is irredeemably irrelevant for biologists, but that using it to identify solutions to the problem as biologists see it requires important reinterpretations and adaptations. Eichner and Tschirhart (2007) however eschew the clarification of this point, and their argument is accepted in the ecological literature without a discussion (as illustrated by its extensive mention in Naeem et al. 2009). It is difficult to see why biologists do not assess the relevance of this model more critically. Our interpretation here is that the false transparency of the notion of biodiversity plays a role in the explanation of the existence of this blind spot. Indeed, this false transparency makes it look as though Eichner and Tschirhart (2007)'s is self-evidently relevant, since it claims to be about biodiversity. We cannot overemphasize that, obviously enough, we do not claim that the term "biodiversity" is the unique, or even the main, culprit in failures of ecological-economic studies of fisheries. The precautions articulated above when analyzing the former case study apply here as well. Our point is more modestly that, by granting a key-role in the coordination between disciplinary approaches to a vague term like "biodiversity", one tends in all likelihood to render invisible the fact that different disciplinary approaches are anchored in different moral assumptions. Like in our former case study once again, we do not claim to unfold a scientific demonstration here, but rather to hypothesize a possible mechanism, accounting for one possible cause among others behind the shortcomings of the models that we analyze.

Eichner and Tschirhart (2007)'s model is just one example, but it is a paradigmatic one, for two reasons.

First, the moral assumption mentioned above is so entrenched in the economic literature that some authors (e.g. Orléan 2011) use it to characterize the vast majority of the current economic literature. This does not mean that economic studies necessarily make this assumption, since heterodox approaches reject it (Lesourne et al. 2006), but rather that this assumption is bound to create recurrent problems in economics/ecology interactions.

Second, the problem witnessed in our example between ecology and economics exists between ecology and other disciplines. For example, numerous anthropological studies presenting themselves as studies of biodiversity acknowledge that they are based on moral postulates (e.g. Mougenot 2003). But if ecologists and anthropologists do not investigate whether their respective moral assumptions are compatible, the possibility for them to provide coherent prescriptions for action is unwarranted.

To sum up the lesson from this third case study: when various disciplines present themselves as studies of different aspects of a common object—biodiversity—they tend to ignore that, if they are based on incompatible moral assumptions—as they often are—the very meaning and usefulness of their interactions are questionable.

17.4 The Way Forward

In the former section, we explored various concrete examples that allowed us to illustrate different kinds of problems created by the false transparency of "biodiversity". The order of presentation was one of increasing complexity and increasing explanatory content: the first example was a simple case of diverging conservation objectives, the second one involved a more interpretative analysis, and the last one eventually allowed us to articulate the crux of the problem created by the false transparency of "biodiversity": collaborative works or disciplinarily studies that conceive of themselves as tackling different aspects of a common object fail to acknowledge the need to ensure that they tackle different parts of a similarly identified problem. They are caged in an illusory shared ontology of the entity biodiversity.

One might argue that the simple solution to all the problems mentioned above would be to get rid of the term "biodiversity" and stop pretending that the various "biodiversity sciences" have anything in common. Such a radical solution (championed, for example, by Santana 2014) could be counterproductive though, by discouraging interdisciplinary collaboration. This would be at odds with the widely accepted idea that interdisciplinary approaches are needed to tackle the globally pressing environmental challenges (Norton 2002; Loreau 2010b).

The aim of the present section is to delineate possible solutions based on the idea that the need to arrive at commonly identified problems should be taken seriously. We first sketch what such a requirement would concretely mean in the case of our three concrete examples, and then we take a broader view.

17.4.1 Facing the Issue of Problem Identification

A leitmotif for contemporary decision analysts, especially those working in multi-actor settings, is that decision-makers requesting decision-support often do not precisely know for what problem they really need support (Bouyssou et al. 2000; Tsoukias et al. 2013). For example, private firms can be aware that they have a problem in their production process because the output is lower than expected. But they don't know if the problem is that they are inefficient or that they were unrealistic when setting their objectives or that their overall conception of what they aim to do was flawed, etc. They know that they have a problem, but can only articulate a rough, ambiguous formulation of it. More interestingly, various stakeholders, for example involved in the management of a complex system such as a watershed, may have only a very partial understanding of the problem that that are nonetheless in charge of tackling. Clarifying a better formulation of the problem, eliminating ambiguities and vagueness often associated with the terms spontaneously used to request decision support, is accordingly a first, critical step for decision analysis (Belton et al. 1997; Rosenhead 2001). Given the nature of the problems identified in this chapter, a similar clarification, on a case-by-case basis, of the precise nature of the problem for the resolution of which (often interdisciplinary) interactions are put to use, may substantially improve the situation.

In the case of the Bel-Air valley, instead of resting content with the fact that they both manage their respective areas according to a scheme that mentions the preservation of biodiversity as an overarching aim, the two managers should answer the following questions. "What is the precise nature of the functional links between the Charente water meadows and the habitats of its tributary, the Bel-Air?", "What are the functional consequences of the absence of a control of the crayfish and Japanese knotweed populations in the Bel-Air on the ecological functioning of the nearby water meadows?", "Would it be justified that the manager of the water meadows contribute (through money or workforce) to help the manager of the Bel-Air to implement specific conservation measures?" These are difficult questions, but sweeping them under the carpet by framing the discussions with the vague consensual terms of "biodiversity" does not make them any less urgent.

In the Taravo case, the question "How to manage the river area in such a way as to promote biodiversity?" is meaningless, because, in this case, the two kinds of stakeholders in charge of the site management have distinct concrete objectives and would implement very different and potentially contradictory action. These differences and discrepancies, however, are hidden by the use of the common word "biodiversity". In this case, a clear management policy and scientific knowledge are simply lacking. The very notion that HCI 3280 is protected under European legislation is nonsense so long as there are no scientific answers to the questions: "Is it possible to define this habitat on the basis of other criteria than species lists?" and "Is it possible to preserve this habitat while controlling invasive species at the same time?" These scientific issues are currently not investigated because, mainly due to the fact that problems are formulated in the vague terms of "biodiversity", these genuine, underlying problems do not surface. Similarly, the national and local strategies regularly produced and updated by environmental institutions are of little use if they do not clarify how the various aspects of biodiversity should be ordered when they conflict in a practical management situation. If such a ranking were available, even if scientific studies turned out to demonstrate that it is impossible to define habitat 3280 without referring to invasive species, a management program could be defined for the lower Taravo on an informed, legitimate basis.

Lastly, in the case of economics/ecology interactions, they would gain much transparency if, instead of rehearsing the vulgate of the supposed biodiversity/well-being link, ecology/economics interactions were systematically anchored in a common identification of the answers to the following questions: "For the purpose of a given decision-making on a conservation issue, what kind of economic information is useful?", "Should we take an anti-paternalistic stance like most economists, or should we rather take a more pedagogic stance and admit that ecological knowledge can rightfully be used to improve everyone's decisions?" and "More generally, what kind of prescription for action is legitimate for biodiversity scientists to formulate on the basis of scientific models?"

17.4.2   A Broader View

The above paragraphs might leave the reader somewhat unsatisfied, since we simply spelled out the questions that the actors should ask themselves *in the specific cases we considered*. Isn't it possible to elaborate a more general approach, liable to help solve problems created by the false transparency of "biodiversity" in a more general setting? We believe that it is possible, and here we sketch our proposal.

17.4.2.1   What Do We Need to Define?

A common view, although often implicit, in the literature, is that the definition of the term "biodiversity" has to be an *objectivist definition*. An objectivist definition is a description of the independent, preexisting entity to which the term to be defined refers. For example, an objectivist definition of the term "Mars" is a description enabling to identify the planet to which the term refers—an object independent from and preexisting our specifying that the term "Mars" refers to it. When one claims to define "biodiversity" by specifying preexisting independent objects, properties or processes, one attempts to provide an objectivist definition.

We have argued above that the ambiguity of the current general definition of biodiversity can create damaging problems, but it is unlikely that objectivist definitions can prevent such problems from arising. Our analysis of the case of ecological-economic models rather suggests that, unless it makes an explicit reference to the value-laden aspirations that make sense of the various biodiversity sciences, a definition can hardly be useful to prevent such problems.

We therefore have to carefully examine the reason why we need a definition. We want to make sure that the various approaches gathered under the umbrella of the term "biodiversity" can provide relevant insights to coherently resolve common problems. The term "biodiversity" provides a form of unification between different approaches and disciplines. But this form of unification is defective when it comes to doing justice to this reason, because it covers up misunderstandings between approaches. What we need is another form of unification, liable to prevent such misunderstandings. Our suggestion is that this unification should rather be buttressed on a general definition of *biodiversity practices*, understood as a coherent collaborative effort from various disciplines to tackle commonly identified environmental or conservation problems.

Our suggestion is therefore to shift the focus from the definition of biodiversity conceived as a putative entity to the definition of biodiversity practices, emphasizing the value-laden aspirations underlying them. This suggestion might seem odd at first sight, because it looks as though one needs to have a prior concept of biodiversity in order to talk about "biodiversity practices". Underlying our suggestion is the idea that such a criticism stems from a linguistic illusion. Our language treats "biodiversity" as a substantive, which makes it look as though "biodiversity practices" necessarily are practices towards the entity to which the substantive refers. But we have argued that there is no such entity biodiversity to which "biodiversity" refers.

Whereas our language gives the false impression that one cannot understand what biodiversity practices are without antecedently understanding what biodiversity is, our suggestion is that the reverse is the case: in order to understand what biodiversity is, what we have to do is to start by thinking through what biodiversity practices are (Sarkar's (2005) approach is very similar to ours in this respect; for an analysis of the differences, see Meinard 2017b).

The search for such a definition of biodiversity practices is bound to be unavoidably largely tentative and interpretative, but it would be misleading to presuppose that this creates a serious problem. The reason is that, when one defines a practice, the suitable definitions cannot be definitive objectivist definitions, because the very act of defining can modify the practice, and this modification can in turn modify the definition. In such a case, the definition does not identify an independent, preexisting practice: it rather participates in the construction of the practice.

The vast literature on the definition of the terms "art" and "artistic practices" perfectly illustrates this idea. Although everybody intuitively knows what these terms mean, there is a vast literature striving to capture definitions of these terms. Unlike biodiversity scientists, art theoreticians have never accepted to rest content with the apparent self-evidence of the central notions of their field: they have endlessly kept trying to find better definitions. It turns out that, in so doing, they have greatly contributed to the enrichment of artistic practices. Indeed, the various definitions of art by prominent art theoreticians have aroused creative responses by artists, who have (more or less consciously) modified their artistic practices to highlight the restrictiveness of these definitions or to explore the avenues they had opened up (Pignocchi 2012).

We argue that, as biodiversity scientists, we should follow this example of art theoreticians. We should always include an explicit definition of the global value-laden biodiversity practices into which we see our studies as being embedded, in such a way as to dissipate misunderstandings like the one highlighted in Sect. 17.3. The point of such references is not just to harbor values, but to prevent misunderstandings. In particular, the value-laden features mentioned in such definitions must be the ones that are crucial to the identification of the general problems that biodiversity sciences should be devoted to solve. If such definitions were systematically formulated, this would launch a creative process by which other biodiversity scientists would modify their practices to criticize the shortcomings or exploit the strengths of each definition, and in turn suggest new definitions, etc.

17.4.2.2 A Tentative Definition

Let us exemplify our approach by articulating our own tentative definition of biodiversity practices:

> Biodiversity practices are studies, actions, strategies based on the aspiration that the development and diffusion of ecological knowledge can lead people to improve their course of action by developing responsible, informed and long-term decision strategies and preferences, mindful of the environmental constraints.

This definition is not the result of a grand deductive, philosophical or scientific, reasoning. It is a tentative interpretation of the studies, actions, strategies that our own experience as conservation scientists and practitioners allowed us to experience—and that our case studies above exemplified. This definition is obviously neither definitive nor uncontroversial. In some contexts, it might appear to be too vague and in need of qualifications or discussions, and the emergence of misunderstandings in the future may require reformulations. But as it stands, it is the kind of definition needed to clarify misunderstandings like the one unveiled above.

For example, if Naeem et al. (2009) had started by articulating such a definition, based on the identification of the problems tackled by biologists, they would most probably have faced difficulties to encompass Eichner and Tschirhart (2007)'s model in it, because these authors do not understand the problem of biodiversity management in the same way as most biologists do. Naeem et al. (2009) would accordingly have admitted the necessity to critically scrutinize the relevance of this model for conservation and ecological purposes. A fruitful critical discussion could have ensued and damaging misunderstanding could have been possibly dissipated.

One seeming problem with this approach is that it is likely that the problems tackled by biodiversity sciences will change over time. Encapsulating them in a single definition of biodiversity practices might accordingly risk encouraging immobility. The tentative definition of biodiversity practices just introduced is, however, liable to play a clarifying role without encouraging immobility because it harbors two crucial features. These two features characterize what we will term a "dynamic definition".

First, since it is granted the status of a tentative definition, it is open to discussion and accordingly flexible enough to continuously adapt to new insights and developments.

Second, since it is meant to be used to critically assess the relevance of various studies for one another, the very act, by a given scientist, to formulate such a definition and test it on a given study can lead him to modify his own practice instead of rejecting the study he assesses. Defining a practice can thereby lead to a modification of this very practice, and this modification can in turn modify the definition.

In this dynamic approach, the best thing that can happen now to the tentative definition of biodiversity practices introduced above is that it be taken seriously enough by biodiversity scientists for them to criticize it, thereby launching the co-evolution of biodiversity practices and their definition.

17.5   Conclusion

The term "biodiversity" is diversely understood by various users, and its general definition is vague. Here we have taken advantage of several case studies to show that this vagueness, which is usually taken by biologists to be innocuous at a theoretical level, can create problems at the concrete level of practical interactions between various approaches to biodiversity issues. The problems studied here share a common structure. In these various settings, the term "biodiversity" is used by various actors to link their respective approaches. The resulting impermeable division of labor, based on the formal but illusory agreement on the objective to study or conserve "biodiversity", hides the fact that the various approaches can promote mutually incompatible goals, eventually leading in conservation practice to self-defeating actions. To end such deadlocks, we have claimed that a clarification, on a case-by-case basis, of the precise nature of the problems for the resolution of which interdisciplinary interactions are put to use is a critical step. It can make the various misunderstandings and contradictions currently impeding management practices due to the false transparency of "biodiversity" visible and subsequently help to resolve them. This case-by-case approach then allowed us to develop a more general proposal, delineating a path towards the resolution of problems created by the false transparency of "biodiversity". The logic of this path can be summed up in four steps:

1. There is need to clarify a general definition of *biodiversity practices*, understood as a coherent collaborative effort from various disciplines to tackle environmental or conservation problems commonly identified on the basis of coherent value-laden aspirations.
2. General definitions of biodiversity practices are always tentative, because the very act of defining them can lead to a modification of our theoretical and practical approaches to biodiversity theorizing and management.
3. In our contributions, we should all make it a rule to always define the global value-laden biodiversity practices into which we see our studies as being embedded, in such a way as to prevent, as far as possible, misunderstandings with other biodiversity studies or actions.
4. We should seize every opportunity to discuss and criticize the definitions put forward by the other biodiversity scientists who have followed the steps above.

Although they have never formally articulated it, art theoreticians and artists have historically followed a similar path, which proved to be very fruitful. Our hope is that biodiversity scientists can learn from this example.

Acknowledgements This work was supported by the Fondation pour la Recherche sur la Biodiversité. We wish to thank E. Casetta, D. Flynn, A. Krajewski, L. Lhoutellier, J. Marques da Silva, X. Morin and D. Vecchi for their comments and corrections.

References

Angeler, D. G., Sanchez-Carrillo, S., Garcia, G., & Alvarez-Cobelas, M. (2001). Influence of Procambarus clarkii (Cambaridae, Decapoda) on water quality and sediment characteristics in a Spanish floodplain wetland. *Hydrobiologica, 464*, 89–98.
Belton, V., Ackermann, F., & Shepherd, I. (1997). Integrated support from problem structuring through alternative evaluation using COPE and V-I-S-A. *Journal of Multi-Criteria Decision Analysis, 6*, 115–130.

Beninde, J., Fischer, M. L., Hochkirch, A., & Zink, A. (2015). Ambitious advances of the European Union in the legislation of invasive alien species. *Conservation Letters, 8*, 199–205.

Bensettiti, F. (Ed.). (2001–2005). *Cahiers d'habitats Natura 2000. Connaissance et gestion des habitats et des espèces d'intérêt communautaire (7 volumes)*. Paris: La Documentation française.

Bouyssou, D., Marchant, T., Pirlot, M., Perny, P., Tsoukias, A., & Vincke, P. (2000). *Evaluation and decision models: A critical perspective*. Dordrecht: Kluwer Academic Publishers.

Brussard, P. F., & Tull, J. C. (2007). Conservation biology and four types of advocacy. *Conservation Biology, 21*, 21–24. https://doi.org/10.1111/j.1523-1739.2006.00640.x.

Burch-Brown, J., & Archer, A. (2017). In defense of biodiversity. *Biology and Philosophy, 32*(6), 969–997. https://doi.org/10.1007/s10539-017-9587-x.

Burgman, M., Carr, A., Godden, L., Gregory, R., McBride, M., Flander, L., & Maguire, L. (2011). Redefining expertise and improving ecological judgment. *Conservation Letters, 4*, 81–87.

Correira, A. M. (2001). Seasonal and interspecific evaluation of predation by mammals and birds on the introduced red swamp crayfish Procambarus clarkia in a freshwater marsh (Portugal). *Journal of Zoology, 255*, 533–541.

DeLong, D. C. (1996). Defining biodiversity. *Wildlife Society Bulletin, 24*(4), 738–749.

Eichner, T., & Tschirhart, J. (2007). Efficient ecosystem services and naturalness in an ecological/economic model. *Environmental and Resource Economics, 37*, 733–755.

ETC-BD. (2015). *Habitat Directive European article 17 database*. European Topic Center on Biological Diversity. www.eea.europa.eu/data-and-maps/data/article-17-database-habitats-directive-92-43-eec-1. Accessed 8 Sept 2018.

European Commission. (1992). Council Directive 92/43/EEC of 21 May 1992 on the conservation of natural habitats and of wild fauna and flora, O.J. L206, 22.7.1992, pp. 7–50. eur-lex.europa.eu/legal-content/EN/TXT/?uri=CELEX:31992L0043

European Commission. (2013). Interpretation manual of European Union habitats. EUR 28. ec.europa.eu/environment/nature/legislation/habitatsdirective/docs/Int_Manual_EU28.pdf

European Union. (2015). *The state of nature in the European Union*. ec.europa.eu/environment/nature/pdf/state_of_nature_en.pdf. Accessed 8 Sept 2018.

Evans, D., & Arvela, M. (2011). *Assessment and reporting under Article 17 of the habitats Directive – Explanatory note and guidelines for the period 2007–2012*. Final Draft. CTE/BD. circabc.europa.eu/sd/d/2c12cea2-f827-4bdb-bb56-3731c9fd8b40/Art17%20-%20Guidelines-final.pdf. Accessed 2 April 2016.

Faith, D. P. (1992). Conservation evaluation and phylogenetic diversity. *Biological Conservation, 61*, 1–10.

Fleishman, E., Noss, R. F., & Noon, B. R. (2006). Utility and limitations of species richness metrics for conservation planning. *Ecological Indicators, 6*, 543–553.

Gaston, K. J., & Spicer, J. I. (2004). *Biodiversity: An introduction* (2nd ed.). Malden: Blackwell.

Gereco. (2014). *Etude floristique de propriétés en espace naturel sensible de la Charente-Maritime*. Unpublished report.

Gereco. (2015). *Elaboration de cartographies de sites Natura 2000 en Corse-du-Sud. Site Nature 2000 Embouchure du Taravo et alentours*. Unpublished report.

Gravel, N. (2008). What is diversity? In T. A. Boylan & R. Gekker (Eds.), *Economics, rational choice and normative philosophy* (pp. 15–55). Abingdon: Routledge.

Hausman, D. M., & McPherson, M. S. (2006). *Economic analysis, moral philosophy, and public policy* (2nd ed.). Cambridge, MA: Cambridge University Press.

Jeanmougin, M., Dehais, C., & Meinard, Y. (2016). Mismatch between habitat science and habitat directive: Lessons from the French (counter)example. *Conservation Letters, 10*(5), 635–644.

Josanoff, S. (2012). *Science and public reason*. Abingdon: Routledge.

Knight, A. T., Cowling, R. M., Rouget, M., Balmford, A., Lombard, A. T., & Campbell, B. M. (2008). Knowing but not doing: Selecting priority conservation areas and the research–implementation gap. *Conservation Biology, 22*, 610–617.

Kolm, S.-C. (2005). *Macrojustice*. Cambridge: Cambridge University Press.

Lee, K. N. (1993). *Compass and gyroscope—Integrating science and politics for the environment.* Washington, DC: Island Press.

Lesourne, J., Orlean, A., & Walliser, B. (2006). *Evolutionary microeconomics.* Berlin: Springer.

Lindenia. (2015). *Etude pré-opérationnelle à la restauration, l'entretien, la gestion et la mise en valeur du Taravo. Phase 3. Programme pluriannuel de gestion.* Unpublished report.

Loreau, M. (2010). *From populations to ecosystems.* Princeton: Princeton University Press.

Loreau, M. (2010b). *The challenges of biodiversity sciences.* Oldendorf/Luhe: International Ecology Institute.

Mace, G. M., Norris, K., & Fitter, A. H. (2012). Biodiversity and ecosystem services: A multilayered relationship. *Trends in Ecology & Evolution, 27,* 19–26.

Maclaurin, J., & Sterelny, K. (2008). *What is biodiversity?* Chicago: The University of Chicago Press.

Maguran, A. E., & McGill, B. J. (Eds.). (2011). *Biological diversity. Frontiers in measurement and assessment.* Oxford: Oxford University Press.

Maris, V. (2010). *Philosophie de la biodiversité.* Paris: Buchet Chastel.

Maris, V., & Béchet, A. (2010). From adaptive management to adjustive management. *Conservation Biology, 24*(4), 966–973.

Mason, N. W. H., et al. (2003). An index of functional diversity. *Journal of Vegetation Science, 14,* 571–578.

MEA. (2005). *Ecosystems and human well-being: Biodiversity synthesis.* World Resources Institute. https://www.millenniumassessment.org/documents/document.354.aspx.pdf. Accessed 8 Sept 2018.

Meinard, Y. (2011). *L'expérience de la biodiversité.* Paris: Hermann.

Meinard, Y. (2017a). La biodiversité comme thème de philosophie économique. In G. Campagnolo & J.-S. Gharbi (Eds.), *Philosophie Economique* (pp. 319–346). Paris: Matériologiques.

Meinard, Y. (2017b). What is a legitimate conservation policy. *Biological Conservation, 213,* 115–123.

Meinard, Y., Coq, S., & Schmid, B. (2014). A constructivist approach toward a general definition of biodiversity. *Ethics, Policy & Environment, 17,* 88–104.

Meinard, Y., Dereniowska, M., & Gharbi, J.-S. (2016). The ethical stakes in monetary valuation for conservation purposes. *Biological Conservation, 199,* 67–74.

Moran, D. (1994). Contingent valuation and biodiversity: Measuring the user surplus of Kenyan protected areas. *Biodiversity and Conservation, 3,* 663–684.

Mougenot, C. (2003). *Prendre soin de la nature ordinaire.* Paris: Édition de la Maison des Sciences de l'Homme.

Naeem, S., Bunker, D. E., Hector, A., Loreau, M., & Perrings, C. (Eds.). (2009). *Biodiversity, ecosystem functioning, and human wellbeing.* Oxford: Oxford University Press.

Nehring, K., & Puppe, C. (2002). A theory of diversity. *Econometrica, 70,* 1155–1198.

Norton, B. G. (2002). *Searching for sustainability.* Cambridge: Cambridge University Press.

Norton, B. G. (2005). *Sustainability.* Chicago: The University of Chicago Press.

Orléan, A. (2011). *L'empire de la valeur.* Paris: Seuil.

Petchey, O. L., & Gaston, K. J. (2002). Functional diversity (fd), species richness, and community composition. *Ecology Letters, 5,* 402–411.

Pignocchi, A. (2012). *L'œuvre d'art et ses intentions.* Paris: Odile Jacob.

Purvis, A., & Hector, A. (2000). Getting the measure of biodiversity. *Nature, 405,* 207–219.

Rodriguez, C. F., Becares, E., & Fernandez-Alaez, C. (2005). Loss of biodiversity and degradation of wetlands as result of introducing exotic crayfish. *Biological Invasions, 7,* 75–82.

Rosenhead, J. (2001). *Rational analysis of a problematic world* (2nd rev. ed.). Wiley: New York.

Sagoff, M. (2008). *The economy of earth* (2nd ed.). Cambridge: Cambridge University Press.

Santana, C. (2014). Save the planet: Eliminate biodiversity. *Biology and Philosophy, 29,* 761–780.

Sarkar, S. (2005). *Biodiversity and environmental philosophy.* Cambridge: Cambridge University Press.

Scharks, T., & Masuda, Y. J. (2016). Don't discount economic valuation for conservation. *Conservation Letters, 9*(1), 3–4.

Sen, A. K. (2002). *Rationality and freedom*. Cambridge, MA: Harvard University Press.

Tsoukias, A., Montibeller, G., Lucertini, G., & Belton, V. (2013). Policy analytics: An agenda for research and practice. *EURO Journal on Decision Processes, 1*, 115–134.

United Nations. (2013) *Convention on biological diversity*. Rio De Janeiro.

Weitzman, M. L. (1992). On diversity. *The Quaterly Journal of Economics, 107*, 363–405.

18

Scientistic, Normativist and Eliminativist Approaches to Biodiversity

Sahotra Sarkar

Abstract This paper argues that biodiversity should be understood as a normative concept constrained by a set of adequacy conditions that reflect scientific explications of diversity. That there is a normative aspect to biodiversity has long been recognized by environmental philosophers though there is no consensus on the question of what, precisely, biodiversity is supposed to be. There is also disagreement amongst these philosophers as well as amongst conservationists about whether the operative norms should view biodiversity as a global heritage or as embodying local values. After critically analyzing and rejecting the first alternative, this paper gives precedence to local values in defining biodiversity but then notes many problems associated with this move. The adequacy conditions to constrain all natural features from being dubbed as biodiversity include a restriction to biotic elements, attention to variability, and to taxonomic spread, as well as measurability. The biotic elements could be taxa, community types, or even non-standard land cover units such as sacred groves. This approach to biodiversity is intended to explicate its use within the conservation sciences which is the context in which the concept (and term) was first introduced in the late 1980s. It differs from approaches that also attempt to capture the co-option of the term in other fields such as systematics.

18.1 Introduction

Many commentators have noted that the term "biodiversity" is of very recent vintage even though biodiversity conservation has become one of the best-known components of both popular and technical discussions of environmental goals today (Takacs 1996; Sarkar 2005, 2017a). The term and associated concept(s) were only

S. Sarkar (✉)
University of Texas, Austin, TX, USA

Presidency University, Kolkata, India
e-mail: sarkar@austin.utexas.edu

introduced in the context of the institutional establishment of conservation biology as an academic discipline in the late 1980s. The introduction of the term is usually attributed to Walter G. Rosen at some point during the organization of a 1986 National Forum on BioDiversity held under the auspices of the United States National Academy of Sciences and the Smithsonian Institution (Takacs 1996; Sarkar 2002).

Originally "biodiversity" was only intended as a shorthand for "biological diversity"; by the time the proceedings of the forum were published as an edited book (Wilson 1988), the new term had been promoted to become its title. The BioDiversity forum was held shortly after the founding of the U.S. Society for Conservation Biology in 1985 (Sarkar 2002). Soulé's (1985) manifesto for the new discipline of conservation biology and Janzen's (1986) exhortation to tropical ecologists to undertake the political activism necessary for conservation had appeared in the previous two years. A sociologically synergistic interaction between the use of "biodiversity" and the growth of conservation biology as a discipline then occurred and it led to a reconfiguration of environmental studies so that the conservation of biodiversity became a central concern. Conservation biology, starting in the 1990s, was conceptualized as the goal-oriented discipline devoted to the protection of biodiversity. Soulé (1985) drew a powerful analogy between conservation biology and medicine; biodiversity was the analog of health.

The existence of a goal engenders a corresponding norm for evaluating whether an action contributes to that goal and, in many contexts, of assaying the extent to which it does so. All the major programs for biodiversity conservation, *viz.*, conservation biology (Soulé 1985), conservation science (Kareiva and Marvier 2012), and systematic conservation planning (Margules and Sarkar 2007), acknowledge the normative component of biodiversity conservation. Not surprisingly, many environmental philosophers have followed suit in treating biodiversity as at least partly a normative concept (Callicott et al. 1999; Norton 2008; Sarkar 2008, 2012b).

But not all. Some philosophers (e.g., Maclaurin and Sterelny 2008), following the lead of many biologists (see Gaston 1996b and Takacs 1996), have treated biodiversity as if it were a purely scientific concept bereft of normative content. That perspective has led to a wide variety of scientific (more accurately, *scientistic*) definitions of biodiversity, each disputed, and with no prospect of resolution of these disputes. The persistence of these disputes has led to many deflationary accounts of "biodiversity" (e.g., Sarkar 2002) as well as proposals to eliminate the term completely (e.g., Morar et al. 2015; Santana 2017). These varied approaches have recently been reviewed by Sarkar (2017a) and that discussion is very briefly summarized in Sect. 18.2 of this paper.

Section 18.3 turns to the core purpose of this paper: a defense of normativism in defining biodiversity. Any such defense must address the question: whose norms? Global norms invoked by Northern conservationists must be pitted against the local norms of communities whose livelihoods are often threatened by biodiversity conservationists' interventions. Section 18.3 traces the ideological underpinnings of global normativism, then rejects it, and critically endorses the use of local values to define biodiversity. But endorsing local values is hardly unproblematic. Section 18.4 examines the problems that beset local normativism.

Accepting normativism does not mean rejecting the use of science any more in biodiversity conservation than it does in healthcare practices. For biodiversity, a partial synthesis is possible. Section 18.5 argues that a rich tradition of discussions within biology of what constitutes biodiversity can be used to lay down adequacy conditions that constrain the latitude available to normative definitions of biodiversity. It also lays out how this synthetic proposal, integrating values and (ostensibly value-free) technical science, can be used in the practice of conservation. Section 18.6 consists of some final remarks.

18.2 Approaches

Sarkar (2017a) has recently distinguished four approaches to defining biodiversity:

1. *Scientism*: Definitions falling under this rubric claim to use non-normative criteria to define and quantify biodiversity. Three such criteria have most often been deployed: richness, difference, and rarity. Each criterion has been used not only singly but also in conjunction with the others. Richness, measured by the number of units, is probably what most users of "biodiversity" have in mind when the term is not explicitly defined. It has also been partly or wholly explicitly defended by Gaston (1996a) and Maclaurin and Sterelny (2008). Difference, interpreted as complementarity, or how many new biodiversity units are introduced to those already present in an entity (such as an area or a community), has been contrasted to richness and promoted by proponents of systematic conservation planning (Sarkar 2002; Sarkar and Margules 2002; Margules and Sarkar 2007). Rarity, interpreted as endemism, along with richness has formed the basis for identifying biodiversity hotspots (Myers 1988; Myers et al. 2000).

   The main problem with these attempts, pointed out by critics such as Santana (2017), is that there seems to be no possible potential resolution of the disagreements between proponents of the different scientific definitions of biodiversity. Difficulties abound: for instance, even within ecology it has long been recognized that richness alone cannot be an adequate characterization of diversity because it does not take equitability into account (Sarkar 2007).[1]

   Efforts to decide between scientific definitions of biodiversity inevitably end up requiring the use of extra-scientific criteria. For instance, proponents of complementarity argue that its use is preferable to richness as a characterization of biodiversity because of the following argument: Consider three potential conservation areas, A, B, and C of which only two can be prioritized. Let A have

---

[1] Consider two ecological communities, A and B. Let A consist of 90 % species μ and 10 % species ν. Let B consist of 50 % species μ and 50 % species ν. Both A and B have richness 2 (assuming species are the relevant unit). Yet, there is a clear sense in which B is more diverse than A. Richness does not capture that sense.

richness 100, *B* have richness 90, and *C* have richness 50. If diversity is to be characterized as richness, the diversity ranking of the three areas would be *A* > *B* > *C* and choosing the best two would mean choosing *A* and *B*. However, suppose that *A* and *B* have 80 units in common. Then *A* and *B* together would contain 110 units. Now suppose that *A* and *C* have 30 units in common and *B* and *C* have 5 units in common. Then *A* and *C* would contain 120 units and *B* and *C* would contain 135 units. Thus the richness-based choice of *A* and *B* is the worst choice for biodiversity representation even if we use total richness as the relevant criterion for the biodiversity content of the prioritized set of conservation areas! This leads to the principle of complementarity (Vane-Wright et al. 1991; Sarkar 2012b): a new conservation area should be prioritized from the available ones on the basis of how many new units it adds to what is already present in those that have been chosen earlier.[2] The relevant point here is that the argument assumes that only two of the three potential conservation areas can be prioritized. Science does not supply this assumption. Its provenance is the existence of some resource constraint that must be respected.

Consider another choice: should richness or endemism or both be a component of biodiversity? Richness appears natural but, as seen earlier, its use is fraught with problems. How about endemism? We may opt for it out of concern for the rare and unusual. The point, though, is that these are no longer scientific claims. We have moved on to talk about values, what aspect of natural variety we deem most worthy of conservation, that is, there has been a transition to an analysis of norms. These cases are typical: extra-scientific considerations are necessary to adjudicate between conflicting scientific definitions of biodiversity.

2. *Eliminativism*: The failure of scientism in the definitional enterprise has led to one extreme response: proposals to eliminate the use of the term "biodiversity" altogether. Such a position has been forcefully argued by Morar et al. (2015) and Santana (2017). However, such a response would only become plausible if there is no other alternative to scientism. The rest of this paper argues that there is a plenitude of other available options. Suffice it here to note that banning "biodiversity" in current environmental discourse would be a daunting task and require efforts that, presumably, even eliminativists would accept as being better used to ensuring conservation in practice.[3]

3. *Deflationism*: Eliminativism as a response to the failure of scientism was preceded by a weaker strategy of deflationism. A strong form of deflationism was an assumption that, not only was there no fact of the matter about what biodiversity is, but that how it should be defined depends on local contexts, and can be gleaned by studying the practices of conservation biologists, for instance, what

---

[2] Note that this choice does not guarantee that the total richness (that is, the number of unique species) would be maximized. In the example earlier, it would lead to the choice of *A* and *C* rather than *B* and *C*.

[3] For more details of these arguments, see Sarkar (2017a). Meinard, Coq, and Schmid, (Chap. 17, in this volume) give a different perspective on why eliminating "biodiversity" or even allowing it to remain irreducibly vague would lead to problems for the practice of conservation.

is being optimized when areas are prioritized for conservation (Sarkar 2002; Sarkar and Margules 2002).

Strong deflationism was problematic for a variety of reasons, most notably perhaps because it seemed to leave no role for explicit discussion of how biodiversity should be defined, even in a given context. It was replaced by a weaker form in which normative discussion of what merits conservation determines what constitutes biodiversity (Sarkar 2008). But this takes us to normativism.

4. *Normativism*: Normativism will be developed in some detail in Sect. 18.3. What motivates this set of definitions is the recognition that the preservation of natural variety is a desirable social goal. For more than a generation, environmental ethicists have argued about the proper warrant for the admissibility of such a goal without reaching consensus (Norton 1987; Sarkar 2005) but, as environmental pragmatists have argued (e.g., Minteer and Manning 1999), these intractable foundational disputes are almost always beside the point in the practical contexts that determine how a conservation policy is formulated and whether it succeeds or fails. For environmental pragmatists, what is of paramount importance is achieving agreement on practical courses of action, shelving foundational disputes in favor of policy achievement. What matters in such contexts is to map, evaluate, and critically engage the values of legitimate decision-makers. These values are not determined by scientific inferences drawn from biological data though those data may—and should—inform the values of the decision-makers. What is critical is a community's vision of the future it desires including but not limited to its perception of its proper role in the natural world. Natural variety is one of those values and the one that is reflected in *biodiversity*; but biodiversity need not be the only natural value. Given this motivation, it remains to develop normativism more systematically. That discussion begins by moving beyond these assertions to arguments designed to establish that biodiversity must be a normative concept. In line with environmental pragmatism, there will be no further attention to foundationalist concerns in this paper.

18.3 Normativism

There are three loosely related arguments that aim to show why biodiversity must be a normative concept. To motivate these arguments consider what is perhaps the most general scientistic definition: biodiversity is the variety of life at all levels of structural, taxonomic, and functional organization. As Gaston (1996b) has documented, many biologists have defended similar definitions (e.g., McNeely et al. 1990; Wilson 1992; Johnson 1993). Is this what *biodiversity* means? If so, it does not seem plausible that biodiversity is the goal of conservation for at least two reasons: (1) There is the venerable ethical principle, *ought implies can*. Can all of biodiversity as defined above be conserved? Ecological communities left undisturbed lose species diversity through competitive exclusion. Evolving populations lose genetic (that is, allelic) diversity through natural selection. Conserving all such diversity is in practice impossible; (2) Is all such diversity in principle a desirable target of conservation? The human skin hosts thousands of microbial species though interpersonal variability is not as high as in the gut which hosts millions (Grice et al. 2009). Should we feel an imperative to conserve all the microbial diversity on the human skin or gut? Bacterial pathogens are rapidly evolving diversity to generate resistance in response to innovation in antibiotics designed to contain them. Other pathogens have shown similar, if less spectacular, responses to drugs. Should such diversity also merit active conservation?

The first argument for normativism begins with the assumption that concepts should be understood against the historical context of their introduction and use.[4] For biodiversity that context is the establishment and institutionalization of conservation biology as an academic discipline. As noted earlier, programs for conservation have always accepted the goal-orientation of the project, and the existence of that goal endows biodiversity with an irreducibly normative aspect. Proponents of conservation biology from the 1980s fundamentally disagree about goals with proponents of systematic conservation planning from the 2000s and, especially, the new conservation science from the 2010s (Kloor 2015) but they all agree with the goal-orientation of conservation. In most cultural contexts, pathogen variability is seen as removed from "biodiversity" with its attendant positive connotation.

The second argument builds on the first. As a result of the goal-orientation of conservation, biodiversity has always been used with a positive connotation. It consists of those aspects of biotic variety that should be conserved. That does not necessarily include all of natural variety. Though the rhetoric of contemporary political discourse often suggests otherwise, not all diversity is positive (Sarkar 2010). A society with extreme economic disparities is more diverse than one that is more egalitarian; but it certainly is not better. A population with both healthy and sick individuals is more diverse but less desirable than one that has only healthy individuals.

The third argument notes that, by the time "biodiversity" was introduced in the 1980s, there had been a generation-long tradition of defining and studying diversity within ecology (Sarkar 2007). Much of this work was spurred by a central theoretical hypothesis of ecology dating back to the 1950s, that diversity begets stability of ecosystems. While both the empirical and theoretical status of this claim continues to be debated today, by the mid-1980s its exploration had led to the formulation of a large variety of diversity (as well as stability) measures. These measures and the

---

[4] This claim is open to philosophical dispute: for instance, adherents of a hard distinction between the context of discovery and context of justification, etc., may deny this assumption (perhaps most famously developed by Mach in his study of physical concepts in the late nineteenth century). Those who view science through the lens of analytic metaphysics and study concepts through intuition and abstraction may also deny it. These issues will be left for another occasion. Suffice it here to note that core analytic methodologies of concept formation (for example, Carnapian explication) accept the relevance of the pragmatic context of conceptual innovation (Carnap 1950).

associated concepts they were supposed to quantify, in contrast to biodiversity, did not display normativity in their use. It is telling that this body of work was entirely ignored by conservation biologists attempting to define biodiversity in the 1990s and since. The most plausible interpretation of this lack of interest in the existing work on ecological diversity is that they viewed their own normative enterprise of designating aspects of natural variety for protection as distinct from these earlier ecological efforts. Thus, scientism was irrelevant to that enterprise. But, then, what requires explanation is why the explicit statements of definitions of biodiversity from biologists, as recorded by Gaston (1996b) and Takacs (1996), are almost always scientistic. Perhaps the explanation lies in the discomfort scientists often feel about explicit normative discussions—but this suggestive explanation is no more than sociological speculation at this point (but see Wolpe 2017).

18.3.1   Global Heritage

For biodiversity, who should set the relevant norms? In the present context this questions amount to asking who determines what aspects of natural variety should be protected. Here conservation efforts have been marred by serious ethical problems reflecting the structural inequities between the global North and the South. Conservation biology was first academically institutionalized in the United States and its agenda reflected the agenda of what has forcefully been criticized from the South as "radical American environmentalism" (Guha 1989). Soulé and his immediate followers had no hesitation in importing their values to the South, at one point arguing that the U.S. federal legal restrictions be circumvented to allow purchase of land for conservation in the South (Soulé and Kohm 1989): "Land acquisition is a very specific need ... The National Science Foundation should view land purchase and maintenance in exactly the same way that it views the purchase of a piece of fancy machinery ... *If there are legal barriers to direct acquisition of land in other countries by U.S. government agencies, then alternatives such as grants to such countries for the establishment and management of research reserves should be explored* " (p. 89; emphasis added). Available aid money would be better spent satisfying the desires of conservation biologists than, for instance, improving livelihoods of local people: "A potential funding source would be Public Law 480 programs which are currently operating in many developing countries" (p. 89).

If Soulé's strictures were imperialist proclamations, Janzen (1986) endorsed the missionary position when he urged: "If biologists want a tropics in which to biologize, they are going to have to buy it with care, energy, effort, strategy, tactics, time, and cash. Within the next 10–30 years (depending on where you are), whatever tropical nature has not become embedded in the cultural consciousness of local and distant societies will be obliterated.... We are the generation [that must] devote [its] life to activities that will bring the world to understand that tropical nature is an integral part of human life" (p. 306). Wilson (1992 and elsewhere) joined many others in declaring biodiversity to be a global heritage. The efforts of Northern conservationists were codified in various documents emerging from global agencies, most notably, the 1992 Rio Convention on Biodiversity.

But claims of global heritage require careful analysis and, when required, systematic deconstruction. Beyond bland assertion, what makes some natural feature or cultural artefact a *world* heritage? As we shall see there is no pat comfortable answer. Global heritage claims typically promote intervention by politically powerful external agents on decisions affecting the habitats of local residents who may have no interest in these global concerns. Moreover, these claims may not even be backed by any legitimate tangible material interest of these external agents—think of protecting a historical ruin just because of its age or a tropical rainforest because of its species richness.

The salience of these issues is borne out by looking at some particular cases: Was it wrong for the Taliban to destroy the Buddhas of Bamiyan? If so, why? And who decides? What gives the so-called international community—which is hardly a community of equals—a legitimate basis for questioning what a community in Afghanistan decides to do with some cultural artefacts present in its domain through no choice on its part? There is no reason to doubt that the strong feelings generated by the destruction of these statues probably reflects some defensible trans-cultural values. But what are they? How can they be spelt out and legitimized? How do these values serve the interest of the international community? Why do these interests override those of the local community? These questions have not received the attention they deserve. To return to the concern of this paper, turn to a biodiversity-related analog (Bevis 1995): Was it wrong for the Malaysians to log the lowland rainforests of Borneo? Why? And who decides? And so on. In this case there is an additional level of complexity. By and large, the local communities in Borneo were resistant to logging (Bevis 1995). The Malaysians opting for development were mainly economic and political elites from the mainland with the required power. The so-called international conservation community, largely activists from Europe and Australia, adopted and possibly manipulated the communities' concerns. But no one bothered to spell out whose heritage the great forests of Borneo were. And why. No matter how strongly we feel about these cases, the answers are not obvious.

Scholars have argued that concepts of heritage emerged in Europe in synchrony with the emergence of nation-states. Meskell (2014) puts it: "Intimately connected with the Enlightenment project, the formation of national identity relied on a coherent national heritage that might be marshaled to fend off the counter claims of other groups and nations" (p. 218). By the nineteenth century, in the late colonial context, the concept of heritage had begun to be applied across national boundaries, especially into the colonies. However, a concept of supranational cultural heritage only began to be formulated after World War I with tentative attempts at its legal codification originally under the auspices of the League of Nations (Boes 2013; Gfeller and Eisenberg 2016).

Full-fledged self-conscious efforts for global heritage designation and protection began with the post- World War II onset of the decolonization era and the formation of the United Nations Educational, Scientific and Cultural Organization (UNESCO) in 1945 (Gfeller and Eisenberg 2016). Claims and designations of global heritage emerged in tandem for both cultural artefacts and natural features. Arguably, especially through the Northern domination of UNESCO and other global agencies, they served to maintain Northern control of these entities in the post-colonial South even after decolonization had brought direct control to an end, for instance, when UNESCO's director Julian Huxley proposed setting aside large areas of central and east Africa as reserves (Huxley 1961; see Adams and McShane 1992 for a critique). (There will be other African examples below.) What is striking is that, beyond implicit appeals to claims of importance for some supranational group of individuals, no argument was advanced to codify why some feature is a global rather than, say, a national heritage; this is a problem that has only recently begun to receive attention (Di Giovine 2015). Instead of argument, attributions of global heritage status have systematically relied on bold assertions by proponents and demands for acquiescence on the part of those who may otherwise have resisted the globalization of their resources.

The first campaign to draw transnational attention to an ostensibly global heritage feature focused on Egypt, starting in the late 1950s, after President Nasser's modernization plan for the country included construction of the Aswan Dam. The project envisioned the submersion of a large number of historic sites and monuments of the Nile Valley, perhaps most notably the Great Temple of Ramses II at Abu Simbel. The plan generated vocal opposition from archaeologists and historians, mainly from Europe; their rhetoric suggested that Egyptians were not legitimate stakeholders in decisions about the fate of these sites (Boes 2013). Though the nationalization of the Suez Canal and his neutrality in the Cold War hardly made Nasser a popular figure in the West, conservationists were able to co-opt him to their campaign in the late 1950s. In 1960 UNESCO undertook an ambitious rescue project of relocating the monuments at risk to higher elevations. Nasser was applauded for recognizing a "right to heritage."

Parallel to the developments around Aswan, two German environmentalists, the father and son team of Bernhardt and Michael Grzimek initiated a global campaign for designating the Serengeti Plain of Tanganyika as a global heritage and "saving" it through formal protection and exclusion of local human use. The core component of their campaign was the creation of the documentary, *Serengeti Shall Not Die*, in which they explicitly and controversially drew an analogy between African wildlife and European historical monuments.[5] Immensely successful, the documentary transformed discussions of the global status of the natural heritage of the South. To continue with the Aswan parallel, shortly afterwards, and this time in India, conservationists from the North, supported by a local elite consisting largely of hunters, co-opted Prime Minister Indira Gandhi to launch Project Tiger in 1973 (Mountfort 1983) in spite of local problems due to tiger-human conflicts. There will be more on Project Tiger below.

---

[5] The German Filmbewertungsstelle Wiesbaden (FMW) dubbed this an "impermissible equation" (*unerlaubte Gleichsetzung*) and its request for the caption's removal captivated op-ed pages in the Federal Republic of Germany with discussions of censorship—see Boes (2013) for more detail.

The normative claims of conservation biology fall into this tradition and are based on the assumption that biodiversity is a global heritage. That is what makes it possible for Soulé to demand the acquisition of land in the South for the benefit of Northern conservationists. Janzen is gentler: he only wants to proselytize and convert the perceived heathens in the name of the global deity that is biodiversity. Indeed, it is commonplace for Northern conservationists to propose policies for distant lands in the South and to demand action (Dowie 2009).

For instance, in the 1980s the British parliament debated sending British troops to Kenya, Tanzania, and Mozambique to protect elephants (Neumann 2004). In the Central African Republic, in the 1990s, Bruce Hayes (a co-founder of the radical environmental organization, Earth First! in the United States), hired mercenaries to shoot at alleged poachers with no semblance of a trial, let alone a fair trial (Neumann 2004). Even when military threats are not used—unlike these African examples— economic power is often deployed against people living near or below the subsistence level if they do not conform to the demands of Northern conservationists (Dowie 2009).

To drive home the point being made, consider a hypothetical example originally constructed by Sarkar and Montoya (2011). Central Texas is home to a suite of endangered and endemic species including birds, salamanders, and arthropods (Beatley 1994; Beatley et al. 1995). In central Texas, attempts to list species under the U. S. Endangered Species Act (ESA), and then to delineate critical habitat and develop habitat conservation plans (as required by law) have long been controversial and have often led to ugly confrontations between landowners and conservationists (Mann and Plummer 1995). Now, imagine that an environmentalist from Mongolia decides to come to Texas, claim expertise on desert landscapes and cave ecology (perhaps justifiably), and demand that prime real estate around the capital city of Austin be converted into a national park. It is intriguing to speculate on the reactions from gun-toting Texans.

But, is there a salient ethical difference between this hypothetical situation and the one in which Oates (1999) (among others) demands more and better-policed national parks in west Africa? Or is it simply a question of power relations? From an ethical perspective, in both situations either we are denying the legitimacy of local sovereignty over resources or we are not. We are either accepting the legitimacy of local residents on the use of habitat or not. If we are forced to conclude that all that differentiates the two situations are power relations, Northern conservationism, as argued earlier, are continuing colonial attitudes and policies in the South (see, also, Guha 1997).

The critical normative issue here is that of parity. What one community—whether it be Northern conservationists or Mongolian desert experts—values should not be transferred without consent to the habitats of other communities. When we couple this normative claim with the realization that a definition of biodiversity is context-dependent in the sense that the valuation of biological resources varies over space (Escobar 1996), then we must turn to local values.

18.3.2 Local Values

Recall that normativism views biodiversity as consisting of entities that merit protection. What is most relevant to the present discussion is that, in practice, different
groups have made different choices (Margules and Sarkar 2007). Let us begin with
governmental agencies and the big non-governmental organizations (derisively
dubbed "BINGOs" by Dowie (2009)) that dominate large-scale biodiversity conservation efforts.[6] In the United States, most governmental agencies adopt endangered
and threatened species as biodiversity units but that is because much of conservation
policy is set in the context of the legal requirements of the Endangered Species Act
(ESA) of 1973. The ESA envisions protection of both animals and plants, includes
subspecies under its purview, but excludes "pest" insects. In contrast, The Nature
Conservancy (TNC), one of the best-known BINGOs, uses habitat types defined by
characteristic ecological communities. Conservation International (CI), another
BINGO, uses both globally threatened and geographically concentrated species.

Some such choice is necessary in order to provide the minimal precision
required to devise conservation policy. Each of these choices reflects cultural
values. For instance, US governmental agencies and CI implicitly presume that
species are the bearers of value. Moreover, they implicitly presume that the
extinction of every species that is admissible (excluding insect pest species) is
equally (normatively) undesirable. TNC implicitly presumes that ecological
communities are the bearers of value. The point is that these definitions embody
cultural norms even though they are often presented as if they are universal and
purely scientific definitions (Sarkar 2008).

Moreover, there are many other equally defensible choices. Sacred groves are
widespread in South Asia, especially in the Western Ghats with evergreen wet
forests and northeastern India, in the Eastern Himalayas. Forest communities of the
Eastern Himalayas have maintained intact patches of cloud forest amidst an almost
completely denuded landscape and have done so in spite of loss of most cultural
associations with their sacred groves due to massive conversion of local populations
to various Christian denominations starting in the mid-nineteenth century. In the
state of Meghalaya, in many of these sacred groves not even deadwood can be
removed.[7] The extant 29 sacred groves occupy over 25,000 ha. These are evergreen
forests on a landscape dominated by limestone. Much of the ecology of the region
continues to be devastated through coal mining and quarrying for limestone besides
swidden farming that has an increasingly shorter cycle (five years now compared to
30 years in 1900). Traditionally each village had at least one sacred grove but local
traditions were largely destroyed by the Christian missionaries. Not one of the
sacred groves has been systematically inventoried except for major tree species; but
they are known to be particularly rich in amphibian species that have a high degree
of microendemism. At least 18 IUCN Red List amphibian species occur in this

---

[6] For more detail and documentation of these examples, see Sarkar (2012b).

[7] Details are from Malhotra et al. (2007) and personal fieldwork.

region. Cave invertebrates in the many caves and fissures under the ground have not been inventoried at all.

Some of the best-known sacred groves of Meghalaya are in the Khasi Hills near the town of Sohra (formerly known as Cherrapunji) which, with an average annual rainfall of 11,430 mm, is one of the wettest places on Earth. (The honor of being the wettest place in the world now belongs to nearby Mawsynram.) Most groves are small and occur on the top of hills but the larger ones also include valleys and the streams that run through them. The most impressive grove here is at Mawphlang which is protected because it is supposed to be inhabited by the spirit "U Basa." Its 80 ha contains at least 400 tree species; the fauna have never been inventoried. The protection regime (known as "Kw'Law Lyngdoh") is severe: Mawphlang is one of the sacred groves from which even deadwood removal is not permitted. The land around the grove is severely degraded.

The complete protection of entire ecological communities may be uncommon even though sacred groves occur throughout the South, especially in sub-Saharan African countries, most notably Ghana and Kenya. In most African countries, sacred groves target a single species or small set of species. Many cultures around the world value individual species in other ways (e.g., as totemic species) that may be of symbolic value or associated with religious practices. Some communities value entire forests. Vermuelen and Koziell (2002) report the case of the Irula hunter-gatherers, a semi-nomadic tribe from Tamil Nadu state of southern India. The tribe is well-known for its association with snakes, both in catching them and in treating snakebites. What this community values is reflected in how they choose a site for settlement. First, they assay a forest for medicinal plants, then snakes, then animals hunted for food or money (rats, rabbits, mongoose, wild cats, etc.). The assessment is complex. The size of animal populations matters and is assessed using the density of footprints. Ecological associations between vegetation type and animals are taken into account (for example, rabbits with *arugampul* (*Cynodon dactylon*), that is, Bermuda grass which, despite its name, originated in West Asia). Typically, in a twist opposite to conventional ecology, animals are taken as indicators for plants. The persistence of forests is critical to the survival of the Irula way of life.

However, this divergence of values need not lead to a vapid cultural relativism in which anything can count as biodiversity. We leave ample room for disagreement which may potentially be resolved: for instance, within a culture we may debate what we value most, whether we value every endangered species as much as we value selected endemic or charismatic ones (species of symbolic and other cultural value). Moreover, cultural values evolve and there can be crosscultural dialectics of engagement, disagreement, and change. Moreover, as we shall see in Sect. 18.5, we may adopt adequacy conditions that delimit which forms of valuing natural entities may count as valuing biodiversity. As an example, if we impose a condition that an adequate definition must value entities that cover a large portion of the taxonomic spectrum, valuing totemic species would not count as valuing biodiversity (Sarkar 2012b). These adequacy conditions will allow a partial synthesis of scientific insight and local values. But science will play a subsidiary role: even these adequacy conditions have to be culturally debated.

18.4   Problems with Local Normativism

Since a form of local normativism is being endorsed here, it behooves us to recognize and pay particular attention to potential problems. There are at least four of these.

18.4.1   Problems of Scale

The last section contrasted local values with global heritage claims about biodiversity. The designation "global" is clear enough in most contexts, referring to Earth as a whole. But "local" is far from clear: it could vary from a community defined by a municipality (or perhaps an even smaller spatial unit) to a nation-state. (Nation-states, in turn, can vary in size from the Vatican with a population of a few hundred or Lichtenstein with a few ten thousand, to China or India each with over a billion.) A few nation-states are ethnically almost homogeneous; while some cities alone embrace scores of culturally distinctive ethnic groups. Is there a natural scale at which biodiversity should be defined or at which conservation measures enacted? The former seems implausible and the latter, as we shall see below, is problematic.

As if to mimic this problem, biotic features that are typically held to merit protection also vary in spatial scale (or extent). In central Texas, microendemic salamanders sometimes have their range restricted to a single neighborhood of a city. The Barton Springs Salamander (*Eurycea sosorum*) and the Austin Blind Salamander (*Eurycea waterlooensis*) both have habitat limited to Barton Springs in the middle of the city of Austin. The Devil's Hole Pupfish (*Cyprinodon diabolis*) is endemic to a single cavern-like habitat in Nevada, United States, and has the smallest known habitat of a vertebrate species, just 0.008 ha (or 80 sq. m.) at the surface (Reed and Stockwell 2014). At the other spatial extreme, the endangered tiger (*Panthera tigris*) ranges from South Asia through Southeast Asia to Siberia (with a large gap at present, though not historically, in China) even after it has lost more than 90% of its habitat during the twentieth century. Earlier it was also present in parts of West Asia.

Different cultural concerns and values may be dominant at different spatial scales. In the case of the two salamander species just mentioned, the International Union for the Conservation of Nature and Natural Resources (IUCN) Red List, the global standard for risk designation, identifies them as "Vulnerable" but this designation is largely irrelevant to their future since the IUCN has negligible influence on conservation efforts in the United States. More pertinently, the United

States Fish and Wildlife Service (USFWS) designates them as "Endangered" which affords them protection under the ESA. So does the state of Texas in its own assessment of risk for its native species. Most importantly, the protection of both salamander species has strong support within the city government of Austin and this support gets translated into actions by city agencies to maintain their habitat. The Barton Springs Salamander, in particular, is woven into the fabric of the city's cultural life. For those who view such endangered species as important components of biodiversity, this is a happy situation.

In contrast, the situation with the tiger is much more complicated. Globally, few species have dominated the consciousness of individuals for centuries as the tigers. About 70% of the world's tigers live in India (Gibbens 2017). At the national level, since the 1970s, tiger conservation has been a priority as exemplified by the 1973 launch of Project Tiger. Since 1972, the tiger has been India's National Animal (replacing the Asiatic Lion, *Panthera leo persica*, a subspecies of which the only extant population is also found in one state, Gujarat, in India). At the local level, conservation is not so simple. Tigers, as predators, often target cattle and other economically important domestic animals. They sometimes prey on humans, especially when habitat degradation and conversion, accompanied by a decrease in their non-human diet options, brings them into close contact with humans. In some tiger habitats, such as the mangrove swamps of the Sunderbans in eastern India and Bangladesh, tigers have long been positively embedded into local culture (Montgomery 1995). In many other tiger habitats in South Asia, human-tiger conflicts have led to local hostility (Gadgil and Guha 1995; Gibbens 2017).

For instance, between 2007 and 2014, in an area near the Chitwan National Park in south-central Nepal, local inhabitants intentionally killed four tigers (Dhungana et al. 2016). In India, local attitudes have been further confounded by the forced dislocation of tens of thousands of resident humans (though accurate numbers are hard to come by) during the process of the creation of Tiger Reserves under the auspices of Project Tiger (Sarkar 1999, 2005). It would come as no surprise that tiger conservation may not be welcome for communities living adjacent to tiger habitats. In fact, local resentment in India sometimes allows tiger poachers to hire local villagers to help them successfully evade anti-poaching efforts using local knowledge; there have even been acts of arson against parks and reserves by villagers adversely affected by their establishment under the aegis of Project Tiger (Gadgil and Guha 1995). Local values in many of these villages will likely not enshrine the protection of as hallowed a conservationist icon as the tiger in India. Returning to our definitional project, tigers would not necessarily be enshrined as a component of biodiversity. What is required are negotiations and tradeoffs between conservationists and victims of tiger depredation.

18.4.2   Conflicts Between Hierarchical Levels

The ambiguity of "local" shows the potential for conflicts between entities at different levels of the political (or cultural) hierarchy from communities through cities, districts, provinces, and the nation-state. These conflicts bear upon choices of a place embedded in different levels of this hierarchy. So, a locality is not only accountable to its community or city values, but also to those of the various regions of which it is a part including the nation-state that may well centralize the most relevant power for nature protection. Returning to the problem of tiger conservation in India, local communities suspicious of tiger conservation are typically pitted against conservationists at every other level of government.

The tiger case is hardly unique. In the late 1980s and early 1990s conservation efforts in central Texas were dominated by programs to protect multiple species besides the salamanders mentioned earlier. These included two migratory bird species, the Golden-cheeked Warbler (*Dendroica chrysoparia*) and the Black-capped Vireo (*Vireo atricapilla*) both of which were eventually declared as endangered by the United States Fish and Wildlife Service (Mann and Plummer 1995). Typically, such a declaration must be accompanied by the designation of "Critical Habitat" for the persistence of the species which imposes some limits on habitat use and transformation. Especially in the case of the Warbler, potential designation of Critical Habitat would have affected a wide swath of central and southern Texas. Opposition from ranchers was such that it is believed to have played a role in the defeat of incumbent Democrat Ann Richards to Republican George W. Bush in the gubernatorial election of 1994 (in spite of a promise by USFWS not to designate any Critical Habitat in a forlorn attempt to save the election for Richards). At the height of the conflict, ranchers explicitly promoted the decimation of endangered species. For these ranchers and much of rural Texas from where they came, these species would not form part of natural values that they would have chosen to protect. Yet, many of the same areas have a long history of private conservation of land and wild areas for a variety of reasons including game management for hunting.

18.4.3   Conflicts Between Localities

Conflicts occur not only across levels of a hierarchy in which a place may be embedded but between places across space. Returning to our well-worne case of tiger conservation, efforts at the national level throughout South and Southeast Asia were for a long time in conflict with China (where, perhaps, a few wild tigers persist) because of a demand for tiger body parts in a set of practices dubbed traditional Chinese medicine. In Southeast Asia many local communities (for instance, in Borneo) value their forests which are viewed as cheap sources of timber in neighboring societies such as Japan (Bevis 1995). There are several species that are protected in their home range because they are perceived to be at risk but categorized as undesirable aliens elsewhere (Marchetti and Engstrom 2016).

It is not being suggested here that these conflicts—across geographical scales, within a hierarchy, or across localities—cannot be resolved. Resolution requires tradeoffs between different groups. Because the use of formal techniques for group decision leads to serious paradoxes (such as the Arrow's theorem—see Sarkar (2012b)), the preferred method for resolution requires deliberation, a process that has many other virtues in the resolution of environmental disputes (Norton 1994). However, there remains another problem, very similar to the conflict between places, but not quite identical; it requires cooperation, rather than tradeoffs, between communities across large geographical scales.

18.4.4 Conservation of Processes

When conservation efforts are directed towards landscapes and seascapes, their focus is typically on individual places (conservation areas), that is, culturally embedded areas with significant biodiversity content, though (as noted earlier) care must be taken to accommodate interactions between such localities. However, protecting places in isolation is rarely enough to ensure persistence of biodiversity. Persistence requires the maintenance of biophysical processes and these occur at multiple scales, from local wind-borne pollination and seed dispersal to ocean currents.

Processes themselves that can become the goal of conservation efforts include long-distance animal migrations. The spectacular 10,000-km migration of loggerhead turtles (*Caretta caretta*) between Baja California (Mexico) and Japan is well known (Shanker 2015). The annual migration of Monarch butterflies (*Danaus plexippus*) in North America is perhaps even more impressive. It is the longest insect migration known to science and the problems faced for its conservation exemplify the difficulties of conserving processes.

There are two North American migratory populations, one with habitat largely restricted to the west of the Rockies, mainly in California and adjoining states, and the other migrating from central Mexico to the north of the United States and southern Canada east of the Rockies. There are also several non-migratory populations in Florida, the Caribbean, Latin America, and elsewhere. (This means that an end to the migration phenomenon does not constitute the extinction of the species.)

The western population mainly winters in California but some insects do move further south through Arizona to Sonora in Mexico. During the Spring most individuals move to the north and east of California. The eastern population, once over a billion individuals, overwinters in a dozen or so high altitude oyamel fir forests in the Transvolcanic Belt of central Mexico, covering the trees like carpets.

All these winter roosts occur within a 100 × 100 km square (Brower and Aridjis 2013). In the spring, after a frenzy of mating, the insects fly north to Texas. Most females lay their eggs on Texas milkweeds, typically attached to the underside of a leaf with only one egg per plant. Most of the wintering generation dies.

The eggs hatch into caterpillars that feed exclusively on milkweeds, pupate, and emerge as adult butterflies. (In contrast, adults feed on the nectar of flowers of a wide range of plant species; milkweeds are no longer particularly important.) The new generation hatched in Texas continues the northward journey. The population fans out, covering much of the United States north of Texas and east of the Rockies. Some butterflies probably change course to turn south to Florida to add numbers to a local non-migratory monarch population found in that state. Most continue going north over a third and, sometimes, even a fourth generation. The northern limit of the migration spans the upper Midwest of the United States onwards to Ontario and the southern edge of Canada. Over these three or four generations the butterflies may travel up to almost 5000 km.

The return journey is even more impressive. The last generation produced in the north travels back to the tiny overwintering area in Mexico. The insects sip nectar for fuel along the way, and flying only by day while typically roosting in small groups for the night. How the insects manage to find their oyamel islands is still poorly understood. Each insect must have both a "map" and a "compass" (Agrawal 2017). Here a "map" means that the insect must know where it is: how the monarch does this remains an unsolved problem. Direction is set by a "time-compensated sun compass" by which each insect uses its internal circadian clock to sense the time of day and the position of the sun to orient itself in the correct direction. When the fall migration starts, the preferred direction is south. The compass is reset during the winter; in spring, the preferred direction becomes north.

For the last few decades, biologists have been warning that this process is endangered. (The species itself is not at risk because of the existence of many non-migratory populations.) Because the overwintering population in Mexico is the entire source of the entire northward migratory population in the spring (and, therefore, of the migratory phenomenon itself), trends in its size are directly relevant to the question whether the migration will persist in the future. These overwintering populations numbered 400 million individuals in the early 1990s but only 100 million yearly since 2010, with a historical low of about 35 million in 2013–2014 (Sarkar 2017b). What has caused this decline remains a matter of controversy.

There is some consensus the degradation and disappearance of the wintering habitat in Mexico has contributed to the migratory population decline. For the wintering habitat, Mexican authorities began systematic conservation efforts in 2000, and these now appear promising in spite of past problems (Víctor Sánchez-Cordero, personal communication). Beyond that, two conflicting hypotheses have been suggested though both could be operative. The *milkweed limitation hypothesis* predicts that spring monarch breeding populations before migration are in decline in the midwestern and northeastern United States and southern Canada. The alternative *migration survival hypothesis* proposes that the southward migratory population is suffering excessive mortality on its way south in Texas and northern Mexico (Sarkar 2017b).

If the milkweed limitation hypothesis is correct, conservation measures should be directed to milkweed restoration at the northern end of the migratory range, and many such efforts have been under way for more than a decade though, arguably, little to show in way of results. If the migration survival hypothesis is correct, efforts should be directed to providing food and shelter to the migrating population towards the southern end of the migration. If both are correct, both measures become important.

The salient point here is that maintaining the monarch migration will require collaboration across a continent-sized landscape. It is dissimilar to the case of conflicts between localities discussed earlier only because there is no potential for a solution through tradeoffs. Those who value monarch migration conservation as an important goal have a difficult task: what they are demanding is the value be attached to a process, not an entity, because the monarch as a species is not at risk of pending extinction.

18.5 A Synthetic Proposal

Where does all this leave us? Recall that, at the end of Sect. 18.3, it was noted that adequacy conditions can be adopted to constrain potential definitions of biodiversity based on local norms. It will be taken for granted that what is being targeted for protection is some aspect of nature (operationally distinguished from what are considered cultural features though, this distinction is not always trivial to maintain). The proposed constraints are intended to prevent all such natural targets of protection to be characterized as components of biodiversity, that is, what, elsewhere, I have called biodiversity constituents (Sarkar 2008, 2012b).[8] These adequacy conditions are necessary to distinguish biodiversity as a value from cases such as: what is valued is some magnificent geological formation, the desire to preserve pristine wildernesses[9], the protection of totemic species alone, the targeted protection of charismatic species such as large mammals in eastern and southern Africa, and so on. This is not to suggest that these are not important and culturally salient goals of conservation; biodiversity is not the only feature that deserves protection.[10]

More importantly, these adequacy conditions can be used to incorporate many, though not all, of the intuitions behind the many scientistic attempts to define biodiversity mentioned in Sect. 18.2. This claim will be elaborated below as the four conditions proposed here are discussed in detail. Suffice it here to know that such a

---

[8] Earlier in the literature, these were called "true surrogates" for biodiversity—see Sarkar (2002), Margules and Sarkar (2002), and Margules and Sarkar (2007).

[9] There are serious ethical problems with wilderness preservation (Woods 2001) but what is most important here is that the
    goal of wilderness preservation is not only distinct but also divergent from biodiversity conservation (Sarkar 1999).

[10] See Santana's contribution to this volume to find a similar claim in different terminology.

strategy allows a partial synthesis between the scientistic and normativist approaches, though only partial because only the intuitions behind the scientistic definitions rather than their specifics get incorporated into this strategy.

What requirements should we impose on potential biodiversity constituents? Here, four adequacy conditions will be proposed[11]:

1. *Constituent entities be biotic*: We are proposing a definition of *bio*diversity. This conditions dates back to Sarkar and Margules (2002). It allows biodiversity constituents to be habitat types, taxa, communities, genes, traits, and so on; but it excludes, for instance, physical environmental features such as rock formations or sand dunes. It also excludes human cultural diversity whether or not cultural diversity contributes to the presence or persistence of biodiversity in a given context.

   Nonbiotic features may well be good surrogates for the constituents in conservation planning. For instance, Sarkar et al. (2005) showed that sets of abiotic environmental classes are often adequate surrogates for varied classes of biota (the putative biodiversity constituents), while many authors have argued that sets of taxa are very rarely good surrogates for each other (Margules and Sarkar 2007) even though they continue to be used (Caro 2010). The success of environmental surrogates does not provide any argument that such abiotic features should be considered as components of biodiversity; rather, it shows that they are good *surrogates for biodiversity*.

2. *Emphasis must be on variability of the constituent set*: That is why it is bio*diversity*. The motivation for this criterion is best explained using an example. Neotropical rain forests have played an iconic role in conservationist campaigns since the mid-1980s, their public appeal perhaps best exemplified by Caufield's (1984) haunting account of their disappearance around the world. Yet, neotropical dry forests are far rarer and more threatened (due to ongoing land cover conversion) than rain forests. When neotropical rain forests, which are arguably over-protected in some regions such as Ecuador, are taken to be emblematic of biodiversity at the expense of neotropical dry forests to the extent of being the basis for a characterization of biodiversity, this condition is not met.

   For habitat types this means that attention should not be restricted to some subset and exclude all others entirely when biodiversity is defined. When dealing with taxonomic groups, this condition also suggests that differences at higher taxonomic levels than that of species are more salient than inter-specific differences. To put it another way, a species that is the sole member of a phylum (e.g., the aquatic species, *Trichoplax adhaerens*, the sole member of Placozoa[12])

---

[11] In my own work, these adequacy conditions have evolved over the years due to continued discussion in many forums—see, for example, Sarkar (2008, 2012b). Condition 4 is being proposed here for the first time.

[12] Note that there is some controversy over this uniqueness claim because some taxonomists feel that there is sufficient genetic diversity within this putative species to distinguish it into several morphogenetically very similar species (Voigt et al. 2004).

is more important for conservation than a species which belongs to a genus with thousands of species (e.g., any jewel beetle species of the genus Agrilus).

3. *Embrace taxonomic spread*: It is particularly important that the definition does not by fiat place arbitrary limitations on the taxa permitted to fall under the scope of "biodiversity." This requirement is probably not controversial. Part of the rhetoric of early conservation biology was that there was a need to move beyond charismatic species that had been the traditional foci of conservation campaigns and embrace the full spectrum of life as worthy of preservation. This rhetoric was often matched by the more concrete proposals that emerged from the field. Its sincerity is being accepted here.

An important function of "biodiversity" was to codify this broadening of conservationist intent. It is arguable that not imposing some requirement that is functionally equivalent to the one being proposed here would miss the entire point of why the new term was enthusiastically adopted in the historical context of its introduction.

4. *Biodiversity constituents must be precise enough for their presence and abundance to be measured:* Within conservation biology in the 1990s, one of the motivations for defining biodiversity was to enable its measurement and quantification. For instance, Williams and Humphreys (1994) begin their discussion with two problems that have to be solved: (1) a relatively theoretical one—what is to be measured? and (2) a practical one—can the data "realistically" be collected? So, it seems reasonable to impose a measurability adequacy condition.

Margules and Sarkar (2002) modified Williams and Humphreys' distinction to distinguish between a quantification problem and an estimation problem which together form what they called a biodiversity assessment problem. Solving the former requires the ability to measure biodiversity constituents *in principle*. That is what this condition requires. Solving the latter problem requires the operationalization of biodiversity for various purposes. For instance, in conservation planning, the detailed spatial distributions of thousands of biodiversity units are required as data. For many biodiversity constituents, obtaining such data, even though in principle possible, is not *in practice* reasonable given time and other resource constraints. What must then be found are adequate surrogates (such as the environmental classes discussed earlier) but these are not part of the definition of biodiversity.

It is instructive to analyze which sets of features survive this adequacy test and which ones do not. One standard approach, that biodiversity is all diversity at the level of genes, species, and ecosystems does not—it calls afoul of Condition 4. Sets of all at-risk species survive community; as do sets of habitat types (so long as they are defined, at least in part, using the ecological communities in them) though it is arguable that the first of these satisfies Conditions 2 and 3 only accidentally rather than as a matter of emphasis. (There is no deep reason why at-risk species— or other taxa—should be varied in their content or span much of the taxonomic hierarchy.)

These cases will probably come as no surprise to conservation biologists who embrace a scientistic attitude to biodiversity. In fact, they show how these adequacy criteria help bridge the gap between normativism and scientism. However, the adequacy conditions also admit non-standard collectives as potential constituent sets for biodiversity, for instance, the sacred groves of Meghalaya (India) discussed earlier. Conditions 1 is obviously satisfied. Condition 2 is satisfied because different kinds of forests present in the region can constitute sacred groves and, internally, they exhibit the variability of tropical cloud forests. Condition 3 is satisfied because each sacred grove is viewed as consisting of all biotic features within them. Condition 4 is satisfied because the number and type of sacred groves in any given collection is relatively easily assayed. If the earlier cases show that the adequacy conditions enable the relevance of scientific intuitions, this one shows how local normativism does not lose out. These conditions permit wide cultural divergences about what type of natural variety merits protection.

18.6   Concluding Remarks

The discussion of biodiversity in this paper has presumed the categoricity of its use in conservation biology and, more generally, biodiversity conservation. However, other areas of biology, in particular taxonomy, have also laid claim to the term over the years. How would the definitional strategy proposed here fare in these areas? Not very well, at least in the case of taxonomy. Taxonomy, by its own explicit goals, is fundamentally a descriptive enterprise; though its theoretical structure does embrace some normative issues, these are epistemological rather than axiological as seen, for instance, in the debates over cladistics (Platnick 1978). Normativism, as outlined here, is simply irrelevant to taxonomy though most taxonomists no doubt embrace many of the normative goals of biodiversity conservation.

How should we address the potential dissonance between the strategy for defining biodiversity presented here and the concerns of taxonomy? The answer given here will be cynical and based on sociological speculation that must be tested against data before the answer is deemed plausible. The speculations: Classical taxonomy had been underfunded since the dominance of molecular biology over the life sciences was established in the 1960s. By the time that conservation biology and "biodiversity" came along in the late 1980s, classical taxonomy based on macroscopic organismic rather than molecular traits, was a dying discipline. Taxonomists jumped on the biodiversity bandwagon when it became apparent that conservation was becoming a powerful current within and beyond the environmental movement. There was money for biodiversity inventory and conservation and, by endorsing that locution, taxonomists could lay claim to some of those resources.

To continue with the cynicism: Conservation biology was supposed to be a "crisis discipline" (Soulé 1985) because species were becoming extinct before biologists could even describe, let alone study, them. With respect to description, the problem was presented as a shortage of trained taxonomists available for that task.[13] The solution? More money for taxonomy. In Costa Rica, there were even moves to generate an army of sparsely-trained "parataxonomists," akin to China's barefoot doctors of the Cultural Revolution, with the task of inventory, producing lists of species at individual locations.

Taxonomy obviously does not place any taxonomic limit on what should be described: the more obscure or difficult a group of organisms, the more technical acuity could be deployed in their classification. From this perspective, the operative measure of biodiversity is species richness (or, possibly, richness at some higher taxonomic level). Success in taxonomy is determined in part by the sheer number of taxa that are successfully described. It is perhaps because they take the claims of taxonomists to be as pertinent as those of conservationists that philosophers such as Maclaurin and Sterelny (2008) embrace richness in their account of biodiversity. A major advantage of this approach is simplicity: richness is conceptually easy to grasp and relatively easy to measure in the field. But the earlier discussions in this paper should also underline the problems.

Why reject the salience of taxonomy? It is time to move beyond cynicism and speculation. The point is that the concept of richness was available to taxonomists long before the advent of "biodiversity." Not only did taxonomy not need the new concept, the neologism made no difference to the practice of taxonomy as a discipline. For taxonomists, "biodiversity" was a slogan, a source of resources for their field.

In contrast, conservation biologists required an operationalized concept of biodiversity to assess the extent to which any measure succeeds or fails because the conservation of biodiversity was the explicit goal of the field (Sarkar and Margules 2002). This is the *argument from necessity*. If we also accept that concepts are best understood in the context of their introduction and use, biodiversity must be understood in the context of conservation biology. But even if we do not endorse this argument from genesis, the argument from necessity makes conservation central to the meaning of biodiversity which, given this context, in turn requires a focus on norms and values for its definition.

Acknowledgements For comments on an earlier version, thanks are due to Elena Casetta, Davide Vecchi, and Jorge Marques da Silva.

References

Adams, J. S., & McShane, T. O. (1992). *The myth of wild Africa: Conservation without illusion.* Berkeley: University of California Press.

Agrawal, A. (2017). *Monarchs and milkweed: A migrating butterfly, a poisonous plant, and their remarkable story of coevolution.* Princeton: Princeton University Press.

Beatley, T. (1994). *Habitat conservation planning: Endangered species and urban growth.* Austin: University of Texas Press.

---

[13] Even as late as 2000, Wilson (2000) was making this claim.

Beatley, T., Fries, T. J., & Braun, D. (1995). The Balcones Canyonlands Conservation Plan: A regional, multi-species approach. In Porter, D. R. and Salvesen, D. A. Eds. *Collaborative planning for wetlands and wildlife: Issues and examples* (pp. 7592). Washington, DC: Island Press.

Bevis, W. W. (1995). *Borneo log: The struggle for Sarawak's forests*. Seattle: University of Washington Press.

Boes, T. (2013). Political animals: *Serengeti Shall Not Die* and the cultural heritage of mankind. *German Studies Review, 36*, 41–59.

Brower, L. P., & Aridjis, H. (2013). The winter of the monarch. *New York Time*, 15 March. http://www.nytimes.com/2013/03/16/opinion/the-dying-of-the-monarch-butterflies.html.   Last accessed 31 May 2017.

Callicott, J. B., Crowder, L. B., & Mumford, K. (1999). Current normative concepts in conservation. *Conservation Biology, 13*, 22–35.

Carnap, R. (1950). *Logical foundations of probability*. Chicago: University of Chicago Press.

Caro, T. (2010). *Conservation by proxy: Indicator, umbrella, keystone, flagship, and other surrogate species*. Washington, DC: Island Press.

Caufield, C. (1984). *In the Rainforest: Report from a Strange, Beautiful, Imperiled World*. Chicago: University of Chicago Press.

Dhungana, R., Savini, T., Karki, J. B., & Bumrungsri, S. (2016). Mitigating human-tiger conflict: An assessment of compensation payments and tiger removals in Chitwan National Park, Nepal. *Tropical Conservation Science, 9*, 776–787.

Di Giovine, M. A. (2015). Patrimonial ethics and the field of heritage production. In C. Gnecco & D. Lippert (Eds.), *Ethics and archaeological praxis* (pp. 201–227). New York: Springer.

Dowie, M. (2009). *Conservation refugees: The hundred-year conflict between global conservation and native peoples*. Cambridge, MA: MIT Press.

Escobar, A. (1996). Constructing nature: Elements for a poststructuralist political ecology. In R. Peet & M. Watts (Eds.), *Liberation ecologies: Environment, development, social movements* (pp. 46–68). London: Routledge.

Gadgil, M., & Guha, R. (1995). *Ecology and equity: The use and abuse of nature in contemporary India*. New Delhi: Penguin Books India.

Gaston, K. J. (1996a). Species richness: Measure and measurement. In K. J. Gaston (Ed.), *Biodiversity: A biology of numbers and difference* (pp. 77–113). Oxford: Blackwell.

Gaston, K. J. (1996b). What is biodiversity? In K. J. Gaston (Ed.), *Biodiversity: A biology of numbers and difference* (pp. 1–9). Oxford: Blackwell.

Gfeller, A. E., & Eisenberg, J. (2016). UNESCO and the shaping of global heritage. In P. Duedahl (Ed.), *A History of UNESCO* (pp. 279–299). London: Palgrave Macmillan UK.

Gibbens, S. (2017). Tiger crushed by excavator in horrific end to human-wildlife conflict. *National Geographic*.   http://news.nationalgeographic.com/2017/03/tiger-india-wildlife-human-conflict/. Last accessed 07 Oct 2017.

Grice, E. A., Kong, H. H., Conlan, S., Deming, C. B., Davis, J., Young, A. C., Bouffard, G. G., Blakesley, R. W., Murray, P. R., Green, E. D., & Turner, M. L. (2009). Topographical and temporal diversity of the human skin microbiome. *Science, 324*, 1190–1192.

Guha, R. (1989). Radical American environmentalism and wilderness preservation: A third world critique. *Environmental Ethics, 11*, 71–83.

Guha, R. (1997). The authoritarian biologist and the arrogance of anti-humanism: Wildlife conservation in the third world. *Ecologist, 27*, 14–20.

Huxley, J. S. (1961). *The conservation of wild life and natural habitats in central and east Africa: Report on a mission accomplished for UNESCO July-September 1960*. Paris: UNESCO.

Janzen, D. H. (1986). The future of tropical ecology. *Annual Review of Ecology and Systematics, 17*, 305–324.

Johnson, S. P. (1993). *The earth summit: The United Nations Conference on Environment and Development*. London: Graham and Trotman.

Kareiva, P., & Marvier, M. (2012). What is conservation science? *BioScience, 62*, 962–969.

Kloor, K. (2015). The battle for the soul of conservation science. *Issues in Science and Technology, 31*(2), 74–79.

Maclaurin, J., & Sterelny, K. (2008). *What is biodiversity?* Chicago: University of Chicago Press.
Malhotra, K. C., Gokhale, Y., Chatterjee, S., & Srivastava, S. (2007). *Sacred groves in India: An overview*. New Delhi: Aryan Books International.
Mann, C. C., & Plummer, M. L. (1995). *Noah's choice: The future of endangered species*. New York: Knopf.
Marchetti, M. P., & Engstrom, T. (2016). The conservation paradox of endangered and invasive species. *Conservation Biology, 30*, 434–437.
Margules, C. R., & Sarkar, S. (2007). *Systematic Conservation Planning*. Cambridge, UK: Cambridge University Press.
McNeely, J. A., Miller, K. R., Reid, W. V., Mittermeier, R. A., & Werner, T. B. (1990). *Conserving the world's biodiversity*. Washington, DC: International Union for Conservation of Nature and Natural Resources, World Resources Institute, Conservation International, World Wildlife Fund, and World Bank.
Minteer, B. A., & Manning, R. E. (1999). Pragmatism in environmental ethics: Democracy, pluralism, and the management of nature. *Environmental Ethics, 21*, 191–207.
Meskell, L. (2014). States of conservation: Protection, politics, and pacting within UNESCO's world heritage committee. *Anthropological Quarterly, 87*, 217–243.
Montgomery, S. (1995). *Spell of the tiger: The man-eaters of the Sunderbans*. Boston: Houghton Mifflin.
Morar, N., Toadvine, T., & Bohannan, B. J. M. (2015). Biodiversity at twenty-five years: Revolution or red herring? *Ethics, Policy & Environment, 18*, 16–29.
Mountfort, G. (1983). Project tiger: A review. *Oryx, 17*, 32–33.
Myers, N. (1988). Threatened biotas:"Hot spots" in tropical forests. *Environmentalist, 8*, 187–208.
Myers, N., Mittermeier, R. A., Mittermeier, C. G., Da Fonseca, G. A., & Kent, J. (2000). Biodiversity hotspots for conservation priorities. *Nature, 403*, 853–858.
Neumann, R. P. (2004). Moral and discursive geographies in the war for biodiversity in Africa. *Political Geography, 23*, 813–837.
Norton, B. G. (1987). *Why preserve natural variety?* Princeton: Princeton University Press.
Norton, B. G. (1994). *Toward unity among environmentalists*. New York: Oxford University Press.
Norton, B. G. (2008). Toward a policy-relevant definition of biodiversity. In G. D. Dreyer, G. R. Visgilio, & D. Whitelaw (Eds.), *Saving biological diversity* (pp. 11–20). Berlin: Springer.
Oates, J. F. (1999). Myth and reality in the rain forest: How conservation strategies are failing in West Africa. Berkeley, CA: University of California Press.
Platnick, N. I. (1978). Phylogenetic and cladistic hypotheses: A debate. *Systematic Zoology, 27*, 354–362.
Reed, J. M. & Stockwell, C. A. (2014). Evaluating an icon of population persistence: The Devils Hole pupfish. *Proceedings of the Royal Society of London B. 281*(1794), 20141648.
Santana, C. (2017). Biodiversity eliminativism. In J. Garson, A. Plutynski, & S. Sarkar (Eds.), *Routledge handbook of philosophy of biodiversity* (pp. 86–95). New York: Routledge.
Sarkar, S. (1999). Wilderness preservation and biodiversity conservation—keeping divergent goals distinct. *BioScience, 49*, 405–412.
Sarkar, S. (2002). Defining "biodiversity"; assessing biodiversity. *Monist, 85*, 131–155.
Sarkar, S. (2005). *Biodiversity and environmental philosophy: An introduction*. New York: Cambridge University Press.
Sarkar, S. (2007). From ecological diversity to biodiversity. In D. L. Hull & M. Ruse (Eds.), *Cambridge companion to the philosophy of biology* (pp. 388–409). Cambridge: Cambridge University Press.
Sarkar, S. (2008). Norms and the conservation of biodiversity. *Resonance, 13*, 627–637.
Sarkar, S. (2010). Diversity: A philosophical perspective. *Diversity, 2*, 127–141.
Sarkar, S. (2012a). Complementarity and the selection of nature reserves: Algorithms and the origins of conservation planning, 1980–1995. *Archive for History of Exact Sciences, 66*, 397–426.
Sarkar, S. (2012b). *Environmental philosophy: From theory to practice*. Malden: Wiley.
Sarkar, S. (2017a). Approaches to biodiversity. In J. Garson, A. Plutynski, & S. Sarkar (Eds.),

Routledge handbook of philosophy of biodiversity (pp. 43–55). New York: Routledge.

Sarkar, S. (2017b). What is threatening monarchs? *BioScience, 67,* 1080.

Sarkar, S., Justus, J., Fuller, T., Kelley, C., Garson, J., & Mayfield, M. (2005). Effectiveness of environmental surrogates for the selection of conservation area networks. *Conservation Biology, 19,* 815–825.

Sarkar, S., & Margules, C. R. (2002). Operationalizing biodiversity for conservation planning. *Journal of Biosciences, 27,* S299–S308.

Sarkar, S., & Montoya, M. (2011). Beyond parks and reserves: The ethics and politics of conservation with a case study from Peru´. *Biological Conservation, 144,* 979–988.

Shanker, K. (2015). *From soup to superstar: The story of sea turtle conservation along the Indian coast.* Noida: Harper Litmus.

Soulé, M. E. (1985). What is conservation biology. *BioScience, 35,* 727–734.

Soulé, M. E., & Kohm, K. A. (1989). *Research priorities for conservation biology.* Washington, DC: Island Press.

Takacs, D. (1996). *The Idea of Biodiversity: Philosophies of Paradise.* Baltimore: Johns Hopkins University Press.

Vane-Wright, R. I., Humphries, C. J., & Williams, P. H. (1991). What to protect? Systematics and the agony of choice. *Biological Conservation, 55,* 235–254.

Vermeulen, S., & Koziell, I. (2002). *Integrating global and local values: A review of biodiversity assessment.* London: International Institute for Environment and Development.

Voigt, O., Collins, A. G., Pearse, V. B., Pearse, J. S., Ender, A., Hadrys, H., & Schierwater, B. (2004). Placozoa—no longer a phylum of one. *Current Biology, 14,* R944–R945.

Williams, P. H., & Humphries, C. J. (1994). Biodiversity, taxonomic relatedness, and endemism in conservation. In P. L. Forey, C. J. Humphries, & R. I. Vane-Wright (Eds.), *Systematics and conservation evaluation* (pp. 269–287). Oxford: Clarendon Press.

Wilson, E. O. (Ed.). (1988). *BioDiversity.* Washington, DC: (U.S.) National Academy Press.

Wilson, E. O. (1992). *The diversity of life.* New York: W. W. Norton.

Wilson, E. O. (2000). A global biodiversity map. *Science, 289,* 2279.

Wolpe, P. R. (2017). Why scientists avoid thinking about ethics. In S. J. Armstrong & R. G. Botzler (Eds.), *Animal ethics reader* (pp. 358–362). London: Routledge.

Woods, M. (2001). Wilderness. In D. E. Jamieson (Ed.), *A companion to environmental philosophy* (pp. 349–361). Oxford: Blackwell Publishers.

19

Eliminativist Approach to Biodiversity

Carlos Santana

Abstract The concept of biodiversity, I argue, is poorly suited as an indicator of conservation value. An earlier concept, natural diversity, fits the role better. Natural diversity is broader than biodiversity not only in moving beyond taxonomic categories to encompass other patterns in the tapestry of life, but also in including abiotic, but valuable, aspects of nature. It encompasses, for instance, geological curiosities, natural entities of historical and cultural significance, and parts of nature with unique recreational and aesthetic value. It allows us to capture the idea of a diversity of ecosystem services, many of which are abiotic or have significant abiotic components. I make the case that refocusing conservation science around natural diversity retains many of benefits of using biodiversity as an indicator of value, while avoiding many of biodiversity's shortcomings. In particular, it provides a framework that highlights the conservation value of many biodiversity "coldspots," avoids the injustice of making conservation primarily the responsibility of the global south/developing world, and fits more neatly with the legal and ethical frameworks used to make conservation decisions in the public sphere.

Keywords Biodiversity · Ecosystem services · Environmental science

Summers in the Great Basin Desert of the Western United States are often intolerably hot and dry, but this is mitigated by the fact that in the Great Basin you're never far from a mountain range. One hot summer a decade or so ago, some friends and I escaped the heat by hiking up Mount Timpanogos, home of the only glacier in Utah. A highlight of the trek was taking a shortcut on the descent by sliding down the glacier. Years later I recounted this to a colleague who moved to Utah more recently,

C. Santana (✉)
University of Utah, Salt Lake City, UT, USA
e-mail: c.santana@utah.edu

and she replied, "what glacier?" A quick online search revealed that the glacier has retreated to below the talus and hikers can no longer slide down it, one of many signs that our Great Basin environment is being reshaped by climate change.

Glaciers won't be the only climate losers in Utah. Decades earlier, famed environmentalist and author Edward Abbey wrote about ascending Tukuhnikivats, a mountain near Arches National Monument. He calls the mountain an "island in the desert" and flees to it in the heat of late August (Abbey 2011: 217). Scrambling up the talus below the summit late in his hike, he hears the whistles of pikas, a sound he equates with the experience of reaching the summit (*ibid*: 224). Pikas are rabbit-like creatures which in Utah live only in alpine and subalpine zones. For the pika, the mountain peaks are quite literally islands in the desert, and as the Earth warms and the tree line creeps upward, those islands will become submerged. Soon there may be no more pikas in Utah, and none of us will be able to relive Abbey's famous experience.

These losses due to climate change are obviously losses of some sort of value. I want to probe how we conceptualize that loss of value. In the case of the pika, one ready answer is that if the pika goes locally extinct, we will have lost biodiversity (in the form of species richness). That biodiversity loss encompasses and explains the various ways in which losing the pika is a loss of value, including the inability to relive Abbey's hike the way he experienced it. But what about the glacier which I slid down as a young man? Its loss is also a loss of value, and it's a shame that my present-day students can't recreate the experience I had when I hiked Timpanogos. The loss of the glacier is not a biodiversity loss, but it feels like a loss of much the same sort as the loss of the pika. This similarity suggests a need for a concept that encompasses both sorts of loss.

Moreover, we need a concept that better captures the way in which having alpine islands in the desert is valuable. It isn't merely the way the pikas and the alpine flora contribute to local biodiversity. By providing a contrast to the desert valleys below— in biodiversity, yes, but also in aesthetic experience, in ecosystem services provided, and even in temperature and humidity—the mountain peaks contribute to Utah's *natural diversity*. The loss of our last glacier is a loss of natural diversity, even if it isn't a loss in biodiversity,[1] and that same feature will be true of all sorts of changes in ecological value.

Biodiversity plays a central role in how we measure and discuss value in the conservation context, but it is an inadequate indicator of ecological value. The more inclusive concept of natural diversity avoids most of these shortcomings without bringing significant new baggage of its own. These reasons, I suggest, warrant placing natural diversity in the central conservation role that biodiversity currently occupies. At the very least, entertaining the concept of natural diversity can, by providing

---

[1] It is probable, of course, that the glacier contributes to the diversity of the microbiome. Suppose, however, that the microbial diversity persists under the talus but the glacier remains inaccessible to hikers. This is still a loss of value. Furthermore, even if some microbial diversity is lost, it is implausible that most of the loss of value is explained by the loss of the microbial diversity.

a useful contrast, clarify many of the issues in assessing and characterizing biodiversity that motivate this volume.

19.1 The Shortcomings of Biodiversity

Biodiversity occupies a central place in conservation science as an object of measurement, a basis for decisions in conservation planning, and as the primary conservation objective.[2] In occupying these roles, biodiversity is an organizing concept which focuses and unifies conservation research, and an umbrella concept (Lévêque 1994) which covers a broad array of conservation concerns. By playing this key role biodiversity serves as a representative of *ecological value*, the complete set of environmental goods of any sort—not only the intrinsic value of natural entities, but the economic, aesthetic, cultural, recreational, spiritual, and health amenities they provide (Santana 2016). Ecological value in this broad sense is the grounds for conservation efforts, so as Norton argues, the right definition of biodiversity is one "rich enough to capture all that we mean by, and value in, nature" (2006: 57). By conserving biodiversity, we aim to conserve ecological value of all kinds. In many ways biodiversity is well suited to this task. Ecological value of most kinds tends to depend on the living organisms that compose ecosystems, and thus on biodiversity. Conversely, since biodiversity relies on many ecological factors, including abiotic processes (Noss 1990), biodiversity conservation entails the conservation of other natural goods. Biodiversity can be operationalized in various useful ways, such as counts of richness (number of units), relative abundance, and measures of difference. It can also be assessed at biological levels of various sorts, such as genes, species, and ecosystems. This makes for flexible, scientifically meaningful, and computationally tractable measures that can feed into conservation planning (Sarkar and Margules 2002). Biodiversity also has an inclusive scope, allowing us to argue for the conservation of species and populations which might fall through the cracks were we to prioritize a different indicator of ecological value. For all these reasons, biodiversity makes sense as a conservation target.

The concept of biodiversity has come under scrutiny, however, in large part because of a sizeable gap between biodiversity value and ecological value construed broadly. It's easy, as a working conservationist, to lose sight of the connection between what's being measured and ecological value. For this reason, even some defenders of biodiversity worry, writing "we do not think that measurement strategies in conservation biology have been convincingly connected to wider theories that show the importance of the magnitudes measured" (Maclaurin and Sterelny 2008: 149). The assumption that our measurements of genes, species, higher taxa

---

[2] Biological diversity also plays important roles outside the context of conservation, such as an explanandum in ecology and evolutionary biology. This chapter isn't concerned with these roles, but more narrowly with biodiversity's central role in conservation.

and so on adequately represent those economic, cultural, aesthetic, and other values needs to be called into question.

Many researchers have done just that, calling into question how our measures of biodiversity relate to ecological value. Sarkar (2002; see also his 2008 and 2016), for instance, notes that the standard units of diversity in conservation biology fail to capture important units of ecological value, such as butterfly migration patterns. He argues that we should adopt a highly-flexible, open-ended concept of biodiversity. On Sarkar's picture, selecting the object of measurement (the *true surrogate*) when we assess biodiversity "is not an empirical question; rather it must be settled by convention" (2014: 5). Specifically, each local, conventional definition of biodiversity should be "based on normative considerations" that reflect the individual context and local cultural values (2014: 5–6). In this way, biodiversity measures can be tailored to account for, say, the cultural significance of a non-endangered species such as the Bald Eagle, which was never in danger of extinction but merited costly conservation efforts (2014). In its most extreme form, this deflationary, contextual account of biodiversity would give up on the content of the concept of biodiversity (i.e. that it is about the biota and about diversity) to allow it to encompass the whole range of ecological values. Sarkar's more recent work (2014, 2016) disavows this extreme stance, but the issues which motivated a more deflationary account of biodiversity remain.

Alternatively, Maier (2012) argues in detail that extant defenses of the value of biodiversity all commit fallacies, perhaps most significantly that of conflating the value of biodiversity with the value of other individual entities such as species and ecosystem processes. It is these natural entities that have value, and not the system-level property of being biodiverse. Motivated by similar issues, Santana (2014, 2016) argues that biodiversity is often a poor indicator of others sorts of ecological value. For example, cultural and aesthetic values often attach to places existing in a preferred state, even if that state has lower biodiversity than a possible alternative. Invasive plant species, for instance, can sometimes coexist alongside indigenous plants, meaning that invasions can increase biodiversity (Cleland et al. 2004). But we are still justified, because of our attachment to historical landscapes, in fighting benignly invasive species. To give another example, the units of biodiversity (species, phyla, genes, etc.) are, in measurement practice, interchangeable with other units of the same type. Consequently, biodiversity measures ignore the way in which some units differ significantly in value from others. They ignore, for instance, how a bat species which eats tons of disease-carrying mosquitos has higher ecological value than the mosquito species it eats.[3] Another set of critics, Morar et al., put the

---

[3] True, biodiversity measures can be sensitive to where organisms sit in the trophic network, not only through direct measures of trophic diversity, but also because the secondary effects of biodiversity loss are mediated by the structure of the trophic network (Dunne et al. 2002), and because trophic factors may regulate levels of species diversity (Terborgh 2015).

But my claim here isn't that biodiversity measures are insensitive to the importance of trophic roles to ecosystem function. I'm claiming that important normative considerations (e.g. malaria is value-negative) are largely invisible to biodiversity measures. Because malaria is value-negative, mosquito species which transmit malaria have less value, and bat species which prey on those

issue succinctly: "there are good reasons to doubt whether [biodiversity] provides any guidance for environmental decision-makers or has any clearly established relationship with those aspects of nature about which we care the most" (2015: 16–17).

In addition to worries about the gap between ecological value and biodiversity measures, Morar et al. accuse the concept of biodiversity as presenting a veneer of scientific objectivity while conservation scientists undemocratically impose their own environmental values on policy-making. According to this argument, the normativity of the biodiversity concept means that the policies supported by the techniques of conservation biology are value-laden. Whose values? The scientists' values, since they choose how to define and measure biodiversity. But this isn't transparent to society at large, thus the values of the broader public might not have an appropriate input to conservation decision-making. The focus on biodiversity conservation is, in effect, an injustice through technocracy. For this reason, Kareiva and Marvier (2012) suggest that conservation biology reframe itself as an interdisciplinary conservation science, one which uses social science to better measure anthropocentric ecological value. The original sin of conservation biology, they argue, was its "inattention to human well-being" (2012: 963). If biodiversity is at the heart of what matters, then "the vast majority of people are a threat" to ecological value, rather than among its beneficiaries (ibid). The focus on biodiversity conservation has thus unjustly relegated socially-oriented ecological values to the background.

Mismatch between biodiversity value and ecological value shows up in practice as well as theory, perhaps most notably in how biodiversity conservation is largely a burden on the "Global South." Although conservation biology emerged in wealthy industrialized countries (the "Global North"), biodiversity increases on a latitudinal gradient that peaks at the equator (Hillebrand 2004), meaning that most biodiversity is concentrated in the less-developed tropical and subtropical nations of the Global South. Likewise, the areas identified by conservation biologists as biodiversity hotspots—the places of highest conservation concern—are mostly in South America, Africa, South Asia, and tropical islands (Myers et al. 2000). We hear a lot about saving the rainforests and coral reefs, but not so much about how the American Midwest is no longer a place where "the buffalo roam and the deer and the antelope play," in the words of a nineteenth century folk song. As a result, conservation has focused much more on the Global South, and while biodiversity conservation and economic development are not always in competition (Tallis et al. 2008), there are almost always tradeoffs (McShane et al. 2011). Most importantly, a chief tool of conservation biology is setting aside protected areas (Rands et al. 2010; Miller et al. 2011), which often imposes significant burdens on local people and indigenous groups (Adams et al. 2004). For this reason, socially-oriented environmental researchers have often been at odds with conservation biologists and environmental philosophers who extoll biodiversity value (Miller et al. 2011). The focus on biodi-

---

mosquitos are more valuable. Bats and mosquitos thus exhibit a value differential, one that is explained by factors independent of their relative contributions to biodiversity, and so isn't easy represented by a biodiversity measure.

versity conservation has thus been a double injustice, placing unfair burdens on the
Global South, and ignoring ecological values in the Global North and biodiversity
"cold-spots" more generally.

19.2   Natural Diversity as an Alternative

Deconstruct biodiversity into its two components, *biological* and *diversity*. My proposal is that the chief virtues of treating biodiversity as the primary target of conservation come solely from the *diversity* component. Conversely most of the
problematic features of biodiversity arise from the *biological* component, since it is
the focus on organisms which excludes many sources of value. We should therefore
aim to retain the benefits of the diversity component while mitigating the drawbacks
of the biological component.

   A good candidate for doing so would be to supplant the concept of biodiversity
in conservation concepts with a diversity concept that extends beyond the biological, which I'll call *natural diversity*. The United Nations Convention on Biological
Diversity defines biodiversity as "the variability among living organisms from all
sources[4];" modeled on that definition, we can define natural diversity as 'the variability among natural entities from all sources'. Note that this is a departure from
some previous usage of the term, which has sometimes equated natural diversity
with what we would now call biodiversity (e.g. in Terborgh 1974, or as used in
U.S. Fish & Wildlife Service directive *701 FW 1*). In the broader sense of natural
diversity that I have specified, preserving natural diversity is a better conservation
goal than preserving biodiversity because it retains the benefits of the diversity component, but moves beyond the biological component.

   The chief virtues of biodiversity as an organizing concept in conservation are its
flexibility and its inclusivity. What makes biodiversity so flexible is the number of
ways we can operationalize it: as species richness, as complementarity, as functional diversity, and so on. As a broader concept, natural diversity could be operationalized even more flexibly, for instance in measures of abiotic soil components,
as diversity of human experience of landscapes (measured through psychological or
economic methods), or as measures of geological composition,[5] to give a few examples. Measures such as these might account for how the loss of a glacier is a loss of
ecological value even if there is no corresponding loss of biodiversity. Natural diversity would thus be a more flexible conservation target.

   It would also be more inclusive, for similar reasons. Natural diversity includes
the diversity of living things, but also other forms of diversity. Lakes whose mineral

---

[4] https://www.cbd.int/convention/articles/default.shtml?a=cbd-02

[5] This might draw on extant conceptions of *geodiversity* (Kozłowski 2004; Gray 2004), or some
alternative. Either way, because natural diversity includes cultural, historical, economic, and expe-
riential components (among other things), it is more than just biodiversity supplemented with
geodiversity.

content is too high to support the diversity of species other lakes support (e.g. Mono Lake or the Great Salt Lake) still contribute to natural diversity, partially in virtue of being inhospitable to most forms of biodiversity![6] A barren sandstone cliff which hosts few organisms, and thus contributes little to biodiversity, might support unique climbing routes and thus have high ecological value. The frigid Kola Peninsula in the Russian Arctic won't show up on any maps of biodiversity hotspots, but its unusual mineral assemblages are of great scientific interest. An area held to be uniquely sacred by an indigenous group is naturally diverse for that reason, even if it isn't biodiverse. A waterway which hosts no endangered species may provide a unique transportation corridor for the local population. If our aim is to conserve natural diversity, all these places will rank high in ecological value, but if biodiversity is our primary conservation goal, they might be ignored. Natural diversity is thus a more inclusive concept than biodiversity.

In addition to surpassing biodiversity in the virtues of flexibility and inclusiveness, natural diversity avoids some (but not all) of biodiversity's vices. Most importantly, as the examples in the last paragraph highlight, it is a better indicator of ecological value. Any natural entity of distinctive ecological value is a significant site of natural diversity in virtue of that distinctiveness. This includes entities of distinctive cultural, recreational, scientific, and economic value, even if those entities contribute little to biodiversity. The gap between natural diversity and ecological value is thus much smaller than the gap between biodiversity and ecological value. Some gap will remain, since some ecological value just doesn't fall under the rubric of diversity in any form, as Maier (2012) and Santana (2016) demonstrate. No single concept that is anything less than intolerably vague is likely to encompass everything of ecological value, however, and that natural diversity does better than biodiversity is a strong point in its favor.

Natural diversity also avoids the potentially unjust ramifications of using biodiversity as the central conservation concept. We have no reason to expect that natural diversity hotspots will cluster in the Global South or that any components of natural diversity besides biodiversity increase on a latitudinal gradient. If natural diversity is the more fundamental concept, conservationists should take the loss of visible stars in a European city to be a loss of conservation value commensurable with (though not necessarily equal to) the loss of an insect species in Madagascar. Consider the following contrast: the rowdy Canadian filmmakers who needlessly damaged unusual natural wonders in the U.S., such as the Bonneville Salt Flats and Yellowstone's Grand Prismatic Spring, got off with a small fine and are publicly called "good young men" (Penrod 2016). In South Africa, by contrast, impoverished hunters who kill endangered species are themselves killed by the hundreds and imprisoned by the thousands (Burleigh 2017). Although damaging a tract of salt flat may not be as morally significant as killing a sentient animal, there is clearly something inequitable in how those who harm natural diversity in the developed

---

[6] Uniquely inhospitable environments host unique organisms such as extremophiles, and so contribute to biodiversity as well, but (a) still have comparatively low biodiversity, and (b) their biodiversity value is the lesser part of their value.

world receive a mere slap on the wrist, while those who harm biodiversity in the developing world can pay with their lives. In terms of conservation value, both cases are serious losses of natural diversity and we should expend commensurate efforts to promote conservation in both cases. In fact, given that individual rhinos and elephants are more easily replaced than individual geological oddities, in terms of conservation values[7] the "good young men" may have caused more harm than any individual poacher. Obviously, few conservationists would endorse the murder of South African poachers. But a biodiversity-centered conservation framework does entail that the actions of the poachers are much more serious than the actions of the thoughtless filmmakers. A natural diversity framework, on the other hand, entails that the actions of both poacher and "good young man" are of a type: harm to natural diversity. The threats to natural diversity, like natural diversity itself, are thus globally distributed, and the burdens of conservation are correspondingly placed as much on the shoulders of relatively wealthy First-Worlders as much as they are on the backs of the Global South.

Another advantage of natural diversity as the central value concept in justifying protected areas is that legal frameworks for establishing conservation areas already appeal to something like it. At the very least, the law typically subordinates biodiversity to a broader class of values. In one chapter of her book *Imagining Extinction*, Heise examines how conservation law around the world is not preoccupied with "mainly a matter of counting how many species have been preserved or have died out," but more with fulfilling "the political, social, and cultural purposes to which it links the conservation of biodiversity" (2016: 91–92). For example, German law "protects endangered species for the sake of conserving culturally defined landscapes rather than habitats for the sake of species" (2016: 90). In other words, the law prioritizes a set of landscapes, which is diversity at the level of natural diversity, not biodiversity. In Bolivia, as Heise recounts, biodiversity conservation is legally situated as part of laws situating the Earth itself as a legal subject, and which invoke the cosmologies of indigenous people (2016: 114–116). Biodiversity conservation's ultimate justification in such a system is thus its contribution to a broader category, "the differentiation and variety of the beings that make Mother Earth" (2016: 116), which sounds more like natural diversity. Even in the United States, where the Endangered Species Act is so central to the environmentalist's toolkit, much, perhaps most, of the legal justification for setting aside protected areas comes from the American Antiquities Act of 1906 (Harmon et al. 2006), which justifies protection not of biodiversity, but of "objects of historic or scientific interest." These sorts of objects are cultural, archaeological, and geological features which may not fall under the rubric of biodiversity, but do contribute to natural diversity more broadly. Furthermore, the designation of national and state parks, which are some of the

---

[7] There are non-conservation values at play here as well: large mammals are sentient, and thus hedonistic values matter as well. But poaching is a bad mostly in terms of conservation value—if the suffering inflicted on the hunted animals was the chief concern, poaching would be no worse than the hunting of unprotected species, and much less immoral than eating factory-farmed meat. For present purposes, we thus have no grounds to say that the reckless filmmakers' crime, because they made no living thing suffer, was less serious than poaching.

most significant protected areas, is justified by the unique aesthetic and recreational opportunities they provide more than the species they protect. Across much of the globe, natural diversity better captures the systems of value behind the legal justifications for conservation, much of which either does not invoke biodiversity or places the value of biodiversity subordinate to a broader class of natural values.

The impulse to protect biodiversity is plausibly grounded in a reasonable aversion towards losing unique, rare, unusual, and distinctive parts of our world. In other words, we find real value in diversity. But this is true not just of biological organisms, but of unique, rare, unusual, and distinctive natural goods of all sorts—of natural diversity inclusive of, but going beyond, biodiversity. An ethnic group's ancestral homeland is unique in virtue of that fact, and merits preservation because in its uniqueness it represents natural diversity. A landscape that is a local rarity, like a glacier in Utah, is particularly valuable in virtue of that rarity. An unusual environment, such as undersea thermal vents, is scientifically valuable in part because it is unusual in abiotic as well as biotic composition. These sorts of natural values fall clearly under a notion of diversity, but not under *biological* diversity. The concept of natural diversity thus retains the conceptual benefits of biodiversity, while better fulfilling the role of capturing value in the natural world. Moreover, the political implications of conserving natural diversity are both more commonsensical and less unjust than a focus on conserving biodiversity primarily. Insofar as they are concepts competing for the same conceptual role, natural diversity thus has a clear advantage.

19.3   Operationalizability

Let's consider a possible disadvantage of natural diversity. Biodiversity serves not only as an indicator of ecological value, but as a measurement concept. Conservationists can operationalize biodiversity in various ways, estimate the amount of biodiversity in different areas, and use these estimates as inputs to conservation decision-making. Natural diversity, as a broader, less cohesive concept, might be more resistant to operationalization, and thus less useful as a measurement concept.

I don't think this is the case, in large part because biodiversity is itself extremely resistant to operationalization and measurement. Natural diversity comes down to meaning something close to "all of nature," and it seems almost nonsensical to try to argue that some areas have more nature than others. But as Sarkar insightfully

---

[8] Many writers have observed that the term *biodiversity* is even more vague than "all of biology." Blandin suggests that *biodiversity* is just a new incarnation of *nature*, and has become "aussi indéfinissable que l'est la nature [as indefinable as nature]" (2014:51). Similarly, others have suggested that biodiversity is interpreted flexibly, with each interpreter using the vagueness of the term to

points out, standard definitions of biodiversity equate it to "all of biology" (2002: 137) as well.[8] Morar et al. contend, with supporting citations to a dozen or so conservation biologists and philosophers, that "the widest consensus about biodiversity understood in this broad and all-inclusive sense is that it cannot, as a matter of principle, be quantified, due to its multidimensionality and the lack of commensurability and covariance among its components" (2015: 18). True, biodiversity can be made amenable to precise measurement, but only by ignoring most its components to focus on merely one or two at a time, such as species richness, genes, traits, habitat types and so on. But this gives biodiversity no advantage over natural diversity. While natural diversity, broadly construed, is utterly unquantifiable, there is nothing preventing picking a couple of dimensions at a time for particular purposes and measuring those. We could, for instance, in selecting conservation areas have measures of recreational usage, number of archaeological sites, and biological family richness, and use all three in tandem to determine what areas to prioritize. This leaves out much of natural diversity, but any practical measure of biodiversity is similarly limited. So, the breadth and vagueness of natural diversity is no different in kind than that of biodiversity.

We might worry, however, that the great flexibility available in selecting indicators of natural diversity will lead to inconsistent measures of natural diversity. Again, on this score natural diversity does no worse than biodiversity. With biodiversity measures, "there will always be some way of comparing (say) one wetland to another that will count the first as the more diverse, and another procedure that will reverse the result" (Maclaurin and Sterelny 2008: 133). This inconsistency across different methods of measurement may be a feature and not a bug, however. Sarkar (2002, 2008) argues that the ability to have different measures of biodiversity which yield different results allows us to fit our concept of biodiversity to the conservation priorities of each local situation, which differ from context to context. If this situational flexibility is a beneficial feature of biodiversity measures, then it would also be a feature of natural diversity measures. The difference, of course, would be even greater flexibility with natural diversity, and a greater ability to use measures which closely track non-biotic entities of ecological value. What might have seemed to be an objection to natural diversity—the sheer range of specific ways to quantify it—turns out to be a point in its favor.

---

read into whatever it is they value in nature. Takacs, for instance writes that "[i]n biodiversity, each of us finds a mirror for our most treasured natural images, our most fervent environmental concerns" (1996:81; cited in Morar et al. 2015). And Blondel observes that biodiversity is "coquille vide ou chacun met ce qu'il veut [an empty shell in which each person places whatever they want to see]" (1995: 225). (Thanks to Elena Casetta for help with the French references and translations).

19.4   More Than Ecosystem Diversity

The astute reader might be wondering whether the concept of natural diversity treads new ground. Doesn't *ecosystem diversity*, for instance, already cover the same territory? Ecosystem diversity is often treated as the one of the three main components of biodiversity, with species and genetic diversity being the others (McNeely et al. 1990). As the diversity of habitat types and community structure and composition (Sohier 2007), it seems that ecosystem diversity captures many of the abiotic entities and landscape-level features that I have argued require a move from bio- to natural diversity. I grant that, when taken seriously as a component of biodiversity, ecosystem diversity addresses some of the worries I have raised. But not most of them, since it only values abiotic entities and landscape features *qua* contributors to biotic activity, and not in terms of many other facets of ecological value, such as aesthetic or economic contributions. For this reason, I think ecosystem diversity still falls short.

19.5   A Natural Bridge

The concept of biodiversity, as I've discussed, has come under attack from several fronts. One reaction to these attacks could be to abandon it and move to some other means of representing and measuring ecological value. Leading alternatives are found in the social sciences, particularly economics, and focus on non-market valuation methods.[9] To abandon biodiversity for economic demand values would likely be to push the non-human biota too far into the background. It would also require us to give up much of the valuable research that conservation biologists have conducted, and leave behind useful tools they have developed. In addition, we would be abandoning a concept that has gained political and rhetorical importance. All these considerations imply that the cost of biodiversity eliminativism is very high.

Natural diversity, I am proposing, is a way to move beyond biodiversity in a way that avoids paying much of this cost. Natural diversity can be a bridge between traditional biodiversity-focused conservation and socially-focused environmental planning, because it includes both biodiversity and human-generated values under an umbrella concept. It suggests that, in principle, various sorts of ecological value are commensurable, and thus we can take research on biodiversity conservation and put it in conversation with other conservation strategies and goals. Of course, the extant concept *ecosystem services* (Millenium Ecosystem Assessment 2005) already tries to do this. Biodiversity and ecosystem services, however, are often an awkward fit. One way to include biodiversity in the ecosystem services framework is to just include biodiversity as a final ecosystem service (Mace et al. 2012). But this is ad

---

[9] For an introduction to these methods of environmental valuation, see Champ et al. (2003).

hoc, and doesn't really suggest in what way biodiversity is comparable to other services. In practice, this often means that valuations of ecosystem services, which are most easily quantified in economic terms, will rate the value of biodiversity quite low (Fromm 2000; Heinzerling 2016). Another means of trying to incorporate biodiversity in the ecosystem services approach is to argue that other services are reliant on biodiversity, but the relationship between biodiversity and ecosystem services is anything but straightforward (Srivastava and Vellend 2005; Mace et al. 2012). What natural diversity offers is an alternative, perhaps superior, means of accomplishing the same goal. Shoehorning biodiversity into the ecosystem services approach falls short because it is ad hoc and has no common standard of comparison. But a natural diversity approach would take the extant tools of biodiversity conservation planning and apply them to a broader set of ecological values, in a natural extension of existing conservation biology. In pitching natural diversity, I'm attempting to refocus our conservation thinking on those glaciers, and landscapes, and ancestral homelands, and other natural features that fall out of the conversation when we discuss biodiversity. But I'm raising the possibility of doing so in a way that is an organic expansion of biodiversity thinking and extant conservation practice, rather than leaving it behind as we embrace a broader set of ecological values.

References

Abbey, E. (2011). *Desert solitaire: A season in the wilderness*. Retrieved from http://ebookcentral. proquest.com

Adams, W. M., Aveling, R., Brockington, D., Dickson, B., Elliott, J., Hutton, J., et al. (2004). Biodiversity conservation and the eradication of poverty. *Science, 306*(5699), 1146–1149.

Blandin, P. (2014). « La diversité du vivant avant (et après) la biodiversité: repères historiques et épistémologiques ». In E. Casetta & J. Delord (Eds.), *La biodiversité en question. Enjeux philosophiques, éthiques et scientifiques*, Les Éditions Materiologiques (pp. 31–68).

Blondel J. (1995). *Biogéographie. Approche écologique et évolutive*, Masson

Burleigh, N. (2017, August 8). Kruger Park South Africa: Where black poachers are hunted as much as their prey. *Newsweek*. http://www.newsweek.com/2017/08/18/trophy-hunting-poachers-rhinos-south-africa-647410.html

Champ, P. A., Boyle, K. J., & Brown, T. C. (Eds.). (2003). *A primer on non- market valuation*. Boston, MA: Kluwer Academic Publishers.

Cleland, E. E., Smith, M. D., Andelman, S. J., Bowles, C., Carney, K. M., Claire Horner-Devine, M., ... & Vandermast, D. B. (2004). Invasion in space and time: Non-native species richness and relative abundance respond to interannual variation in productivity and diversity. *Ecology Letters, 7*(10), 947–957.

Dunne, J. A., Williams, R. J., & Martinez, N. D. (2002). Network structure and biodiversity loss in food webs: Robustness increases with connectance. *Ecology Letters, 5*(4), 558–567.

Fromm, O. (2000). Ecological structure and functions of biodiversity as elements of its total economic value. *Environmental and Resource Economics, 16*(3), 303–328.

Gray, M. (2004). *Geodiversity: Valuing and conserving abiotic nature*. Chichester: Wiley.

Harmon, D., McManamon, F. P., & Pitcaithley, D. T. (2006, January). The antiquities act: The first hundred years of a landmark law. *The George Wright Forum, 23*(1), 5–27. George Wright Society.

Heinzerling, L. (2016). Economizing on nature's bounty. In J. Garson, A. Plutynski, & S. Sarkar (Eds.), *The Routledge handbook of philosophy of biodiversity*. London: Taylor & Francis.

Heise, U. K. (2016). *Imagining extinction: The cultural meanings of endangered species*. Chicago: University of Chicago Press.

Hillebrand, H. (2004). On the generality of the latitudinal diversity gradient. *The American Naturalist, 163*(2), 192–211.

Kareiva, P., & Marvier, M. (2012). What is conservation science? *BioScience, 62*(11), 962–969.

Kozłowski, S. (2004). Geodiversity. The concept and scope of geodiversity. *Przegląd Geologiczny, 52*(8/2), 833–883.

Lévêque, C. (1994). Le concept de biodiversité: de nouveaux regards sur la nature. *Natures Sciences Sociétés, 2*(3), 243–254.

Mace, G. M., Norris, K., & Fitter, A. H. (2012). Biodiversity and ecosystem services: A multilayered relationship. *Trends in Ecology & Evolution, 27*(1), 19–26.

Maclaurin, J., & Sterelny, K. (2008). *What is biodiversity?* Chicago: University of Chicago Press.

Maier, D. S. (2012). *What's so good about biodiversity?: A call for better reasoning about nature's value* (Vol. 19). Dordrecht: Springer.

McNeely, J. A., Miller, K. R., Reid, W. V., Mittermeier, R. A., & Werner, T. B. (1990). *Conserving the world's biological diversity*. Washington, DC: IUCN, World Resources Institute, Conservation International, WWFUS and the World Bank.

McShane, T. O., Hirsch, P. D., Trung, T. C., Songorwa, A. N., Kinzig, A., Monteferri, B., et al. (2011). Hard choices: Making trade-offs between biodiversity conservation and human wellbeing. *Biological Conservation, 144*(3), 966–972.

Millennium ecosystem assessment. (2005). *Ecosystems and human wellbeing: A framework for assessment*. Washington, DC: Island Press.

Miller, T. R., Minteer, B. A., & Malan, L. C. (2011). The new conservation debate: The view from practical ethics. *Biological Conservation, 144*(3), 948–957.

Morar, N., Toadvine, T., & Bohannan, B. J. (2015). Biodiversity at twenty-five years: Revolution or red herring? *Ethics, Policy & Environment, 18*(1), 16–29.

Myers, N., Mittermeier, R. A., Mittermeier, C. G., Da Fonseca, G. A., & Kent, J. (2000). Biodiversity hotspots for conservation priorities. *Nature, 403*(6772), 853.

Norton, B. (2006). Toward a policy-relevant definition of biodiversity. In *The endangered species act at thirty* (Vol. 2, pp. 49–58).

Noss, R. F. (1990). Indicators for monitoring biodiversity: A hierarchical approach. *Conservation Biology, 4*(4), 355–364.

Penrod, E. (2016, November 4). Canadian will pay fine for wakeboarding on Utah's Bonneville salt flats. *The Salt Lake tribune*. http://archive.sltrib.com/article.php?id=4546578&itype=CMSID

Rands, M. R., Adams, W. M., Bennun, L., Butchart, S. H., Clements, A., Coomes, D., et al. (2010). Biodiversity conservation: Challenges beyond 2010. *Science, 329*(5997), 1298–1303.

Santana, C. (2014). Save the planet: Eliminate biodiversity. *Biology & Philosophy, 29*(6), 761–780.

Santana, C. (2016). Biodiversity eliminativism. In *The Routledge handbook of philosophy of biodiversity* (pp. 100–109).

Sarkar, S. (2002). Defining "biodiversity"; Assessing biodiversity. *The Monist, 85*(1), 131–155.

Sarkar, S. (2008). Norms and the conservation of biodiversity. *Resonance: Journal of Science Education*, (7), 13.

Sarkar, S. (2014). Biodiversity and systematic conservation planning for the twenty-first century: A philosophical perspective. *Conservation Science, 2*(1).

Sarkar, S. (2016). Approaches to biodiversity. *The Routledge handbook of philosophy of biodiversity*.

Sarkar, S., & Margules, C. (2002). Operationalizing biodiversity for conservation planning. *Journal of Biosciences, 27*(4), 299–308.

Sohier, C. (2007). *Ecosystem diversity*. Available from http://www.coastalwiki.org/wiki/Ecosystem_diversity. Accessed on 12 Nov 2017.

Srivastava, D. S., & Vellend, M. (2005). Biodiversity-ecosystem function research: Is it relevant to conservation? *Annual Review of Ecology Evolution and Systematics, 36*, 267–294.

Takacs, D. (1996). *The idea of biodiversity. Philosophies of paradise*. Baltimore: Johns Hopkins University Press.

Tallis, H., Kareiva, P., Marvier, M., & Chang, A. (2008). An ecosystem services framework to support both practical conservation and economic development. *Proceedings of the National Academy of Sciences, 105*(28), 9457–9464.

Terborgh, J. (1974). Preservation of natural diversity: The problem of extinction prone species. *BioScience, 24*(12), 715–722.

Terborgh, J. W. (2015). Toward a trophic theory of species diversity. *Proceedings of the National Academy of Sciences, 112*(37), 11415–11422.

20

Biodiversity and the Edible Environment

Andrea Borghini

Abstract The green revolution, the biotech revolution, and other major changes in food production, distribution, and consumption have deeply subverted the relationship between humans and food. Such a drastic rupture is forcing a rethinking of that relationship and a careful consideration of which items we shall preserve and why. This essay aims at introducing a philosophical frame for assessing the biodiversity of that portion of the living realm that I call *the edible environment*. With such expression I intend not simply those plants and animals (including in this category, henceforth, also fish and insects) that were domesticated for human consumption, but also the thousands of species that are regularly consumed by some human population and that are regarded to some degree as wild. The visceral, existential, and identity-related relationship that link humans with the edible environment can be regarded as *sui generis* and can constitute a ground for explaining why it should receive a preferential treatment when it comes to preservation, propagation, and development. First of all, I discuss whether we should draw a sharp divide, when it comes to preservation efforts, between *wild* and *domesticated* species (§1); secondly, I assess whether to draw a sharp divide between *natural* and *unnatural* entities, when it comes to measurements and interventions regarding the edible environment (§2); finally, I ask what is the value of biodiversity as far as food is concerned, and how best to preserve and foster it (§3 and §4). The closing section draws some suggestions for future investigations and interventions.

Keywords Food biodiversity · Wild foods · Natural foods · Food ontology

A. Borghini (✉)
Department of Philosophy, University of Milan, Milan, Italy
e-mail: andrea.borghini@unimi.it

20.1   Introduction

The concept of biodiversity is rather unquestionably associated with the idea of untamed forms of life, living entities, or parts of living entities, which developed on Earth independently of or prior to humans. Far more controversial, instead, is whether biodiversity measurements and interventions should take into account also forms of life and living entities that have been to some degree influenced by human activities (cfr. Siipi 2016, section 1, for a comparison of the opposing camps.) Some authors lean towards a very inclusive notion of biodiversity, which virtually leaves out no (actual or possible) form of life, living entity or any part of a living entity. But, no matter where one wishes to draw the line, it seems unfeasible to have a notion of biodiversity that excludes all those entities that have been *in some way or other* influenced by humans. Not only would it currently appear unfeasible to insist on protecting only those entities that are untamed; more importantly, any effort of preservation or development of such entities would by itself undermine their being in some way or other independent of human existence. Hence, any account of bio-diversity seems bound to address the following two questions:

(1) Are there living entities that should be excluded from measurements of biodi-versity as well as from efforts of conservation[1]?
(2) Should the criteria for inclusion in a measurement or intervention be context-dependent or context-independent? For instance, could different criteria be selected depending on circumstances?

Another important outcome of the literature on the concept of biodiversity is that, at a closer look, most accounts of biodiversity reveal a preference towards more famil-iar forms of life.[2] Thus, for instance, preserving the existence of pandas seems a much more important goal than preserving the existence of any species of mollusks that inhabits some remote marine areas, no matter how important such mollusks may be to a certain ecosystem; or, consider the little attention that the preservation of bacteria has received in comparison to animals or plants, which can arguably only in part be justified by the taxonomic challenge of classifying bacteria. It is important to reflect on the reasons that might have supported such preference of certain forms of life over others; are those *good* reasons, that is, reasons that justify keeping such preferences in our accounts of biodiversity? Or, are such preferences biases, which cannot be justified? Hence, the following additional question for any account of biodiversity:

---

[1] In this paper I shall refer to *conservation*, rather than *preservation*, efforts. I do not have in mind such a sharp distinction between the two notions, as established in the classical dispute between Gifford Pinchot and John Muir; at the same time, it seems most appropriate to speak of conserva-tion of ordinary biodiversity, rather than its preservation, because of the active human role not only in establishing and maintaining it, but also in exploiting it for the purposes of – among others – dieting, pleasure, research, and profit.

[2] See Marques da Silva & Casetta, Chap. 9, in this volume.

(3) What are the grounds for preferring certain (possibly more familiar) forms of life over others? For instance, do such preferences rest on efficiency, or perhaps on some emotional or spiritual connection?

In this essay I aim to address the three questions just raised from a particular angle, which has thus far received relatively sparse attention from philosophers. I aim to analyze the value of biodiversity when it comes to that portion of the living realm that I call *the edible environment*. With "edible environment" I intend not simply those plants and animals (including in this category, henceforth, also fish, insects, mushrooms, and some algae) that were domesticated for human consumption, but also the thousands of species that are regularly consumed by some human population and that are regarded to some degree as wild. The edible environment constitutes a particularly significant point of entry into the preferential attitudes that humans bear towards different forms of life. The visceral, existential, and identitarian relationship that humans bear with the edible environment can be regarded as *sui generis* and, as we shall see, can constitute a ground for answering question (3), that is, why the edible environment should receive a preferential treatment when it comes to preservation, propagation, and development. The edible environment is also an intuitive entry point into questions (1) and (2). Are there edible (parts of) living entities that ought not to be included in measurements of biodiversity (e.g., GMOs)? Should measurements and preservation efforts be contextual; for instance, should they tend to clearly demarcate between biodiversity of the edible environment and other forms of biodiversity? Are the criteria employed to account for the biodiversity of the edible environment specific to it? Are they consistent across the board?

In order to investigate questions (1)–(3), in what follows I will take up a number of issues that cut across them. First of all, the issue of whether we should draw a sharp divide, when it comes to conservation efforts, between *wild* and *domesticated* species. I address this in §1. Secondly, we should assess whether to draw a sharp divide between *natural* and *unnatural* entities, when it comes to measurements and interventions regarding the edible environment; this issue will be at the center of §2. Finally, we should ask what is the value of biodiverse foods and how best to preserve and foster it; these two issues will occupy sections §3 and §4, respectively. In §5 I shall return to questions (1)–(3) and suggest some answers.

20.2   Wild and Domesticated Foods

Today the food for sale at any supermarket, deli, or food market in an agriculturally industrialized country such as the United States, Holland, or Japan is a testimony to two kinds of success stories. The first is the story of human attempt to tweak the edible environment to serve human nutritional, economic, and social purposes; call this *the conquer and divide story*. There are a few remarkable facts about the diversity of domesticated species, which reveal the importance of looking at taxonomic levels below species when it comes to domesticated plants and animals (cfr. Especially Diamond 2002: 702). Only 14 out of 148 large terrestrial mammalian were domesticated, and only about 100 plants out of 200,000 candidates. Any of those species is in itself a remarkable success story, featuring the rise of an astonishing number of varieties[3]: – e.g. over 40,000 varieties of beans, over 10,000 varieties of tomatoes, over 8000 varieties of apples, and circa 8000 breeds of animals (for a concise and up to date overview of the diversity of animals that humans consume, cfr. Chemnitz and Becheva 2014: 22–25). But, the first success story tells also of the many ways in which humans managed to cooperate with microscopic organisms such as bacteria, yeasts, and fungi, to preserve, modify, create key staples, including cheese, yoghurt, beer, wine, vinegar, chocolate, coffee, whisky, and hundreds more.

The second, more recent, success story tells of the increasing connection of food production and distribution systems worldwide; call this *the food revolution story*. Characteristic of it is the decline or extinction of thousands of varieties and breeds produced throughout the long path to domestication. For instance, in 2012 the FAO update on the state of livestock biodiversity estimated that circa 2000 of the 8000 animal breeds are at risk of extinction or nearly extinct. Or, to make two examples regarding plants, of the 287 varieties of carrots that humans devised, only 21 are still cultivated; and of the 8000 varieties of apples that we have a trace of, only 800 are still cultivated (cfr. Fromartz 2006, Chapter 1 and Pollan 2001, Chapter 1) If we look at the broad picture, data from The Food and Agriculture Organization of the United Nations (FAO-UN) indicate that since the beginning of last century 75 percent of plant genetic diversity has been lost. To offer some additional examples, "at least one breed of traditional livestock dies out every week in the global context; of the 3831 breeds of cattle, water buffalo, goats, pigs, sheep, horses and donkeys believed to have in this twentieth century, 16% have become extinct and 15% are rare; some 474 of livestock breeds can be regarded as rare, and about 617 have become extinct since 1892" (*Conservation and Sustainable Use of Agricultural Biodiversity* 2003, paper 3, p. 23). Also, over 97% of the varieties of foods sold in 1900 in the United States had disappeared from the market by 1983 (Cfr. Fromartz 2006, Chapter 1). The shrinkage of the number of varieties is principally due to the increased integration of food markets, controlled by fewer and fewer actors at the origin and during distribution, as well as by a growing syncretism and homogeneity within diets across the planet. Within a globalized food market, only a few varieties

---

[3] Some reader may wonder why the data presented in this section regard *varieties* rather than *cultivars*. Although the two concepts are at times used interchangeably (cfr. ICNCP 2009, Chapter 2, Art. 2.2), the technical usage of 'cultivar' picks out a more restrictive taxonomic notion, based on three principles: (i) possession of a distinctive character; (ii) uniformity and stability of such character; (iii) heritability of said character (ICNCP 2009, Chapter 2, Art. 2.3). At the same time, though, "no assemblage of plants can be regarded as a cultivar or Group until its category, name, and circumscription has been published" (ICNCP 2009, Chapter 2, Art. 9, Note 1); yet, many extant and past varieties, that may suitably comprise a cultivar, were never inventoried; thus, in a discussion of the biodiversity of edible plants, it seems most suitable to at least start off by considering all varieties, and then possibly refine the domain by considering the stricter notion of cultivar.

per species tend to be favored, namely those varieties that deliver an economic advantage such as production cost, shelf life, or consumers' appeal.

Much of the discussion concerning food biodiversity has indeed focused on either the conquer and divide story or the food revolution story. It is hard to overestimate the importance of the first story to human evolution. There are still countless details of the processes of domestication of each animal and plant that await to be uncovered, which will shed light over the economic, medical, social, political, and cultural history of humanity (cfr. Wrangham 2009). Equally important is the astonishing shift in food production and consumption, which occurred since the advent of synthetic fertilizers and, more recently, biotechnologies. In the past century, nearly all varieties on the market have been replaced. This leaves us with two major interrogatives: to what extent biodiversity efforts should focus on preserving ancient varieties, and to what extent measurements of biodiversity within the edible environment should include cultivars created by means of techniques such as lab cloning and genetic modification.

As important as they may be, the conquer and divide story and the food revolution story are far from portraying a comprehensive picture of the biodiversity of the edible environment. Indeed, the two stories leave out so-called 'wild' organisms (which can be counted not only in terms of cultivars, varieties or races, but also higher taxa such as species and families), which not only comprise a very significant portion of human diets, but also reveal a continuum between the discussions of prototypical biodiversity conservation targets (e.g. hot spots and endangered species) and conservation targets within the edible environment. By aggregating a number of studies, ethnobotanists estimated that humans have fed themselves off of over 7000 species (Grivetti and Ogle 2000; MEA 2005). Looking at 36 studies in 22 countries of Asia and Africa, Bharucha and Pretty (2010: 2918) estimated that "the mean use of wild foods (discounting country- or continent-wide aggregates) is 90–100 species per place and community group. Individual country estimates can reach 300–800 species (India, Ethiopia, Kenya)." Most importantly to our purposes, in nearly all countries across the globe, with the notable exception of United States, wild species and domesticated species are tended and consumed jointly, and in a number of occasions they are also jointly marketed. To many farmers, the distinction between domesticated and wild species is, indeed, of little significance. At the outset of their paper, Bharucha and Pretty (2010) report the words of a woman farmer, interviewed in Mazhar et al. (2007: 18), who exclaims: "What do you mean by weed? There is nothing like a weed in our agriculture."

Hunting and gathering have coexisted with agriculture in most societies. Both hunting and gathering, when integrated into the dietary routines of a society, require a deep knowledge of the prey, which encourages strategies for favoring the reproduction of animals and plants, possibly favoring desirable traits.[4] For example, a boar hunter may favor the reproduction of certain boar families, which possess certain particularly desirable traits (e.g. size and build); a gatherer of mushrooms may favor the reproduction of certain species in a spot by facilitating or creating specific

---

[4] Cfr. (Kowalsky 2010).

environmental conditions (e.g. humidity, shade, enclosure from the passage of certain animals, selection of surrounding plants). From this perspective, the so-called wild species found within the edible environment are typically far from the most untamed species known on Earth. This is not surprising since eating is a relationship, which in the case in point involves humans and a few thousands species; with time, although humans did not domesticate such species, they (voluntarily or involuntarily) managed them. The study of biodiversity within agriculture should be undertaken alongside with the study of biodiversity within the wild edible environment. As Bharucha and Pretty (2010: 2923) conclude,

> The evidence shows that wild foods provide substantial health and economic benefits to those who depend on them. It is now clear that efforts to conserve biodiversity and preserve traditional food systems and farming practices need to be combined and enhanced.

Another important consideration, which shows how simplistic is the view that draws a strong divide between wild and domesticated species within the edible environment, is that such a view leaves no place for the myriads of microscopic organisms that are essential to human diets worldwide, with virtually no exception. To illustrate the point with an example, it would be unsound to claim that, at origins, humans domesticated *Saccharomyces cerevisiae* and that sourdough is one of the countless outcomes of such domestication process.[5] After all, humans were not even aware of the existence of such a microscopic fungus when they started making use of it to produce bread, beer, chocolate, etc. Rather, sourdough emerged out of a form of cooperation between humans and a variety of fungus, which was not guided by specific species design, but that likely proceeded through trials and errors guided solely by taste and, more broadly, culinary success. Yet, *Saccharomyces cerevisiae* is arguably part of the biodiversity within the edible environment that humans should aim to preserve. Parallel arguments can be developed with respect to progenitors of domesticated plants that are still lingering in the wild: they are especially precious because they typically preserve the widest genetic pool of the taxon. Hence, the discussion of the biodiversity within the edible environment found in virtually any extant human diet should include not only domesticated species and varieties.

The upshot for subsequent discussion is that any assessment of the measurement of the biodiversity found within the edible environment, and of the best means best conserve it, should recognize how variegated are the relationships that humans created with species in the edible environment. The edible environment is constituted by organisms (or parts of organisms) that can hardly be put on a scale with respect to their untamedness – from the wildest to the most domesticated. This complicates a bit the picture when it comes to decide whether to leave out certain items within the edible environment from measurements of biodiversity and efforts of conservation. Is there really a difference between domesticated and wild species, which

---

[5] It may be more plausibly argued, however, that *Saccharomyces cerevisiae* was in some sense *domesticated* by contemporary biotechnology, through the selection of best suited samples and genetic engineering interventions. I shall leave the issue open here.

should be reflected in the study of the biodiversity of the edible environment? What to make of microorganisms? What about bioengineered plants and animals? These questions shall occupy us in the next section.

20.3   Finding Natural Joints

In her assessment of biodiversity with respect to human modified entities, Siipi (2016) distinguishes between three main ways to devise a cutoff point between what should be included in an inventory of biodiversity and what should not. They are respectively based on the (i) history, (ii) properties, and (iii) relations of the entities under consideration. It may be useful to begin by illustrating the three ways.

(i) With respect to the first way, imagine the case of two portabella mushrooms, one of which is grown wild in a forest and one that is induced by a human in a garage; suppose further that the two mushrooms are genetically identical, because the mushroom mycelia of the wild mushroom have been transplanted in a litter in the garage, that they are hardly distinguishable when it comes to taste and nutritional characteristics (their properties), and that they have similar market and culinary value (their relations); nonetheless, since the mushrooms have different histories, which rest on their different contexts of development, the forest-grown is regarded as 'wild' or 'naturally grown' and the other is labeled as 'home-grown'.

(ii) To illustrate the second way, based on properties, imagine two portabella mushrooms grown side by side in a forest (hence having alike histories) and having similar market and culinary values (relations), but possessing quite different nutritional and gustatory properties, due to the malformation of one of them, developed just a few hours before being picked (hence, not really historically based). You can conceive of a classification according to which the malformed mushroom is regarded as 'unnatural' and the other as 'natural,' based on their morphological properties.

(iii) To illustrate the third way, based on relations, imagine two portabella mushrooms grown side by side in a forest and perfectly comparable in terms of size, nutritional and gustatory properties, but ending up in two different markets and, from there, two different restaurants; although alike in terms of origin and properties, one of the mushrooms *belongs* to the market and restaurants where it ends up, being recognized as a 'natural' element within the edible environment of the culinary culture(s) showcased within the market and the restaurant; the other mushroom instead is considered as somewhat foreign to its market, and ends up in a restaurant to be featured as an exotic, 'unnatural' item to be placed alongside the other foods.

In her paper, Siipi distinguishes in total six criteria for telling apart natural from unnatural foods for the purposes of finding a cutoff point between entities that should be relevant for biodiversity measurements and preservation, and those that should not be so regarded. Three criteria are history-based and concern respectively: the organisms that are independent of humans (e.g. a wild herb spontaneously grown in a remote beach), the organisms that are not controlled by humans (e.g. wild blackberries), and the organisms that are regarded as non-artificial (e.g. Golden Delicious apples). Two additional criteria are property-based and regard the foods that are: alike to spontaneously occurring (e.g. two mushrooms, one forest-grown and one garage-grown, which are alike in terms of culinary and nutritional values); alike to possibly existent foods (e.g. seedless grapes, obtained by grafting spontaneously occurring samples of seedless grapes (cfr. Sperber 2007). The sixth and final criterion is relation-based and rests on whether a certain food 'belongs,' or is 'suitable to' a given context (e.g. grapes being unsuitable for the original climate and soil of Central Valley, California, and thus requiring amounts of water, pests, and herbs that are disproportionate).

Siipi's thorough examination of the ways in which a portion of an edible environment may be found to be natural or unnatural, and hence possibly included or excluded in biodiversity measurements and conservation policies, demonstrates the complexity of the matter at stake. To further her analysis, we should first of all avail ourselves of the conclusion reached at the end of the previous section. Concepts of 'wild' and 'domesticated' do little to usefully represent key relationship between humans and parts of the edible environment that are relevant for the purposes of biodiversity; these concepts should be better substituted with specific histories of the relationships between humans and parts of the edible environment, which evidence characteristic traits. But, which traits should matter? A tentative list should include things such as control over reproduction, difficulty of reproduction, potential variability of the desired trait, nutritional properties, gustatory properties, broadly cultural properties, ecological fit, culinary fit … Yet, can an exhaustive list be provided? Interesting in this context is also to recall that focusing on individual species may not be the best manner to proceed in an assessment of biodiversity; rather, we should look at broader networks of biotic and abiotic entities that produce certain foods. Indeed, the production of certain foods requires the employment of additional living organisms; for instance, any peach orchard requires bees for pollination, or any fig orchard requires a specific species of tiny wasps as well as trees that are both female and male, even though the latter produces fruits that are generally not eaten. More generally, it seems best in a number of circumstances to pay attention to ecosystems that deliver foods, rather than to single species within the edible environment; for instance, Vitalini et al. (2009) proposed a EU designation of 'Site of Community Interest,' which would stress indeed the presence of a number of biotic and abiotic conditions that are necessary to sustain portions of the edible environment. Hence, to guarantee the security of the relevant edible plants, measurements of the biodiversity of the edible environment should take into account not simply the (parts of the) species or varieties that we feed off, but also the other biotic and abiotic conditions that are necessary for their survival.

In conclusion, although a number of traits, such as the ones just listed, are arguably most relevant in the majority of contexts, it seems methodologically incorrect to proceed by devising a list that should fit every assessment of biodiversity within an edible environment. In other terms, this discussion suggests that there is no one vantage point from which to assess the naturalness of a (part of an) organism found within the edible environment; this is because naturalness is not a matter of that (part of an) organism being domesticated or wild, but what function it plays in a system of food production, which ecological relationships it bears to other surrounding organisms, as well as what functions it could play in possible edible environments that are considered relevant for the purposes of the assessment. The remaining of the paper will elaborate on this thesis.

20.4   The Values of a Biodiverse Edible Environment

The previous two sections argued that the biodiversity within the edible environment includes a wide array of entities, which can hardly be systematized in a context-independent taxonomy. Recapping the complexity of the domain under examination will be useful to start assessing its multiple dimensions of value.

While a distinction between wild and domesticated species could be defended, based on the degree of human intervention during reproduction and selection, it would be unseemly to claim that all wild species develop(ed) fully independently of human interference. Some developed actually in conjunction with human artifice. For instance, many forms of gathering and hunting do proceed through subtle modifications of the surrounding environments by humans, which facilitate the reproduction and growth of specific populations of the designated species or variety. The spectrum of domesticated species varies significantly, as it includes plants that are reproduced by cutting (e.g. rosemary, strawberries, avocados), plants that are reproduced by grafting (e.g. grapes, most fruit trees), plants that spread by sexual reproduction (e.g. most grains), and a number of plants that can reproduced in any of those ways (e.g. avocados, cacao trees). For any of the domesticated species, we can wonder what is the degree of interference that humans have access to in *any single instance of reproduction*: with animal farming, breeding is often controlled down to the minutest details by the farmer; but, with grains, it is not feasible to control the path of all the pollen in a field and often also the selection of seeds is only partially decided by the farmer, who will work with what was provided by the previous harvest; in an orchard, the farmer cannot control the process of pollination by bees to the minutest details; ditto for controlling reproduction within a fish farm. Alternatively, we can measure the overall degree of interference between humans and the species by looking at the genetic distance of domesticated organisms from their wild progenitors, factoring in the number of generations that occur between the two samples.

Species are not the main units when biodiversity of the edible environment is at issue. Rather, cooperation among different clusters of organisms, organized in more or less spontaneous communities, seem to be the key concept. In this light, the cooptation of microorganisms to produce viable foods, which at least until Pasteur proceeded somewhat blindly with respect to the biological details of the microorganisms, comprises a chapter of its own in the inventory of the relationships between humans and the edible environment. Beer, bread, chocolate, yoghurt, cheese are all examples of a *culture* of fermentation, which played an essential role in food production and conservation and in human evolution. Fermentation, as we know it today, is the outcome of a microbial diversity, which is formed during food preparation and aging, and which confers a distinctive specificity to food; not only it is arguable that, without being fermented, beer would not be *beer*, but it is also arguable that without certain strains and species of microbes that are characteristic to it, a certain beer style (e.g. pilsner) would be not that beer style. The research on this issue is extensive; cfr. Borghini 2014: 1118 and, for a recent significant example based on cacao Ludlow et al. 2016). But, in the human cooptation of certain microorganisms with the aim of producing viable foods, what matters most? Is it most important to preserve – say – the spontaneity of a process (as it happens in spirits which are spontaneously fermented) or, rather, to preserve certain final characteristics of a product? Do (some aspects of) the genetic profiles of the microbes fix the identity of the final product? Or, rather, should a certain style, brand, gustatory profile be privileged? These questions suggest that the diversity within the microbial world correlated to the edible environment is not all on the same level, and that privileging the diversity derived from a type of process (e.g. spontaneous fermentation) may hinder the tending to the diversity of other aspects of the fermentation process, such as the preservation of certain strains or varieties.

Finally, we should consider the complexity of biotic and abiotic factors that favor the reproduction, growth, and development of the edible environment. Hence, species of bees and wasps, varieties of soils, minerals within water, etc. It seems that an inventory of the biodiversity of the edible environment should include these items too, since they are arguably essential to the creation of a number of products, such as geographical indications (see Borghini 2014). Hence, the biodiversity of the edible environment is bound to include also a vast array of entities that are not really edible, but that are conducive to the production of the foods we eat, currently or possibly. There is here an overlap between the concept of biodiversity, when applied to edible items and when applied to non-edible items; for instance, we have reasons to protect the biodiversity of soils for reasons that are independent of food production (cfr. Brussaard et al. 2007), and yet the study of biodiversity of the edible environment will argue for their protection too.

The fluidity between the different categories of entities found within the edible environment is also reflected in the fluidity between roles taken up by food workers. For example, it is common for farmers to act as custodians of both domesticated and wild species, and to be farmers of both agricultural products and weeds; also, gatherers and hunters are oftentimes also farmers; and it is common to a fisherman to hunt too.

The edible environment showcases the complexity of the idea of biodiversity because of the multifarious forms that the relationship between humans and edible items can take and has historically taken. Such complexity is mirrored also in the *reasons* we have for valuing biodiversity within the edible environment. Because the subject matter are ordinary living entities, which are considered in relationship to humans, it is fairly obvious that the reasons to value the biodiversity within the edible environment is entrenched with human existence and culinary cultures. I shall here divide up the field in four points: (i) food sovereignty; (ii) food security; (iii) gastronomic pleasure; and (iv) intrinsic value. Let us illustrate each of them, in order.

(i) *Food sovereignty*. With food sovereignty we intend the ability within a group of people to self-determine a sufficiently ample and relevant portion of their dietary choice by means of food production. Food sovereignty emphasizes, hence, the active ability of a society to determine which plants and animals to harvest and produce, as well as the means of production. Such a power of a society is foundational with respect to the possibility (not the necessity, of course) of actively fostering biodiversity within the edible environment. This power is especially critical when it comes to farming societies that are economically, technologically, and politically at a disadvantage. It was indeed introduced in 2002 by the World Bank with the International Assessment of Agricultural Knowledge, Science and Technology for Development (IAASTD), a 3-year program aimed at improving food production knowledge and technology within disadvantaged societies. But, the idea of food sovereignty was implicitly already at the core of the *Universal Declaration on the Eradication of Hunger and Malnutrition*, issued during the World Food Conference of 17 December 1973. The declaration begins by noting "the grave food crisis that is afflicting the peoples of the developing countries where most of the world's hungry and ill-nourished live and where more than two thirds of the world's population produce about one third of the world's food;" it then goes on to suggest that "all countries, and primarily the highly industrialized countries, should promote the advancement of food production technology and should make all efforts to promote the transfer, adaptation and dissemination of appropriate food production technology for the benefit of the developing countries and, to that end, they should inter alia make all efforts to disseminate the results of their research work to Governments and scientific institutions of developing countries in order to enable them to promote a sustained agricultural development."

(ii) *Food security*. If food sovereignty regards the foods that are produced within a society, food security concerns instead the kinds, qualities, and quantities of foods *accessible for consumption* within a society. A consistent portion of the literature on the biodiversity of edible organisms focused on the importance of an ample spectrum of nutrients for combating malnutrition, when manifested both as a lack of sufficient calories or nutrients (undernutrition), or as an excessive amount of calories or nutrients (overnutrition) (cfr. Borghini 2017 for a philosophical analysis of hunger). In their literature review on food security and biodiversity, Chappell and La Valle (2011) provide significant evidence that "alternative agriculture, which is generally targeted at sustainability and compatibility with biodiversity conservation, is indeed on average better for biodiversity conservation than conventional agriculture, which usually (though not always) targets increases in yields to the exclusion and even detriment of direct concerns about biodiversity, equitability, and food access." (Chappell and La Valle 2011: 17) Chappell and La Valle's conclusion, which stresses the link between food sovereignty and food security (see also Jarosz (2014) on this point), goes hand in hand with the so-called "ecoagriculture approach" (McNeely and Sherr 2002), according to which landscape biodiversity is key to ensure sustainable farming practices that are in sinking with their surrounding ecosystems.

A limitation of much literature on food security and biodiversity rests on a narrow conception of the edible environment, which is basically limited to agricultural products. As we have discussed above, landscapes comprising wild and domesticated foods come into closest contact in some of the regions where food access is most insecure. The availability of a diverse spectrum of plants, animals, and other suitable living entities for setting up an edible environment is a form of empowerment for communities that aim to improve their conditions with respect to food sovereignty (see below) and food security. Farming in urban or rural regions that present adverse climatic conditions or inadequate natural resources can be much improved by a wide stock of living entities that can adapt to different circumstances. Thus, an approach such as ecoagriculture is best appreciated when conjoined with the thesis that there is no sharp discontinuity between wild and domesticated species, and no easy cutoff point between natural and unnatural entities, at least when it comes to the edible environment.

Fostering biodiversity, hence, can aid to food security at two different levels: at the ecosystem level, and at the level of the edible environment. At the ecosystem level, biodiversity can facilitate a sustainable availability of resources, to be employed by producers with the edible environment. At the level of the edible environment, the wider the stock of organisms available to any producer worldwide, the higher will be her power to deliver suitable goods to a market; this, in turn, will increase opportunities for a diverse diet, which is key to address malnutrition. Of course, the availability of certain goods on the market is far from granting, by itself, a solution to food insecurity (cfr. Chappell and La Valle 2011: 17–18, for a discussion of this point); yet it is certainly a necessary step in order to address it.[6]

(iii) *Gastronomic pleasure.* Both food sovereignty and food security are linked to gastronomic pleasure, as they by and large shape the link between dining and civic values (cfr. Alkon and Mares 2012). Promoting a biodiverse edible environment is a mean to empower communities not only by strengthening the

---

[6]I should mention a difficult question arising at this point in connection to bioengineered organisms, which cannot be fully developed here. Consistent efforts are underway to engineer organisms (e.g. GM crops) and foods (e.g. lab meat), which may add to the diversity available to farmers worldwide. Arguably, these items should be included in an inventory of the (actual or potential) biodiversity of the edible environment; but, should they rank as equally valuable as their non-genetic counterparts? In keeping with the approach presented above, an answer to this question can be provided only when faced with a broader decisional framework, which keeps into account also the other three values to be considered next, that is gastronomic pleasure, food sovereignty, and intrinsic value.

sustainability of their agricultural production and by improving the likelihood that they will be food secure, but also by allowing them to diet in a manner that is most in keeping with their ethical, political, religious, an identity-related values. In this sense, the biodiversity of the edible environment is directly linked to the power of a society of choosing and determining its diet.

Slow Food may have been the first and to date the most fervent voice to insist on the link between biodiversity and gastronomic pleasure. The Slow Food movement, founded in 1986 by Carlo Petrini among others, focused since the beginning on the importance of gastronomic pleasure for any conversation concerning the political, ethical, and socio-economic discourse about food. In a telling passage of *Slow Food Nation*, Petrini writes: "Pleasure is a human right because it is physiological; we cannot fail to feel pleasure when we eat. Anyone who eats the food that is available to him, devising the best ways of making it agreeable, feels pleasure." (2013: 50) Now, for Petrini gastronomic pleasure is directly linked to the availability of a diverse array of products, which in turn can be obtained only by actively encouraging the diversity of forms by means of which humans tend the edible environment. Hence, gastronomic pleasure necessarily passes through the promotion of the diversity of the edible environment, by supporting typically small-scale tending practices, which aim to express the most meaningful relationship that humans can establish with the edible environment.

Unthinkingly, it may seem that pleasure is an accessory feature of human relationship to food, which should be kept out of the ethical and political sphere. Nonetheless, in the past three decades it has become increasingly more evident that there is a link between gastronomic pleasure and such issues as biodiversity, malnutrition, food sovereignty, and food access. Petrini's position echoed that of Wendell Berry (cfr. 1990) and has been re-proposed in different forms by several additional authors, such as Pollan (cfr. 2006) and Stiegler (cfr. 2006). Thompson (2015: Chapter 3) especially lays out a convincing discourse showing the link between dieting and the ethico-political sphere. It is impossible to tell apart the meaningfulness of the pleasure experienced during the act of eating and the sorts of food that are consumed (cfr. Borghini 2017 on this point); such pleasures are most often (positively or oppositionally) linked to values imbued in a society, to empowerment, and civic values, no matter how ordinary they may seem in any single dining occasion. For these reasons, gastronomic pleasure is to be included within the spiritual ecosystem services.

(iv) *Intrinsic value.* Finally, the value of a biodiverse edible landscape may rest on the value of the species, the varieties, and the trophic chains themselves. This may be the most intuitive value of a biodiverse edible environment in the context of a general discussion of the philosophy of biodiversity. A wider spectrum of forms of life has not only a utilitarian value, perhaps quantifiable in monetary terms like Costanza et al. (1997) provocatively suggested; rather, it is worth to invest time and resources in the fostering of biodiversity because there is a beauty and value in its mere *existence*, regardless of the consequences. When it comes to edible landscapes, the history of painting offers some neat illustrations of the view of those who hold that biodiversity should be regarded as an intrinsic value. The paintings of Bartolomeo Bimbi, for example *Pears of June and July* (1696), entertain the spectators by simply showcasing a mesmerizing array of pears cultivated under the Medici family at the end of '600 s.[7]

In closing, it is important to ask whether the four reasons for measuring the edible environment are in some way affected or affecting a diet; more specifically, should a diet be influenced by consideration of biodiversity or, vice versa, do dietary decisions influence our stance on the measurements of the biodiversity of edible plants and animals? The fourth reason suggests implicitly that the wider a variety of edible items in a diet, the more commendable the diet; and you may wonder whether such a constraint is acceptable. You can fancy a society that is food secure, sufficiently pleased when it comes to dining, whose members have in some way come to agree in an equitable manner upon their diet, and which nonetheless survives within an extremely monotonous diet (made, perhaps, of one daily pill synthesized in dedicated laboratories). This society would arguably not contribute to the fostering of the biodiversity of the edible environment; should its members still pay dues to those in other societies who, instead, aim to foster it? If they should, in what measure should they contribute? For instance, suppose that the vast majority of the world population would come to prefer such a diet; would it still be feasible to maintain the goal of fostering an edible environment as diverse as we currently have? In other words: does the specific diet undertaken by an individual, or a society, maintain obligations to others who chose different diets? In what measures?

Since the biodiversity of the edible environment depends on human tending possibly in a more active manner than the biodiversity of other forms of life, these questions are far from trivial to answer. An important upshot is that, if we accept that the biodiversity of the edible environment is valuable independently of its consequences, then we should keep tending edible items even if they were to phase out of any human diet. To what extent this is a feasible goal is an issue that is worth further, future investigation.

20.5   What to Foster Within the Edible Environment?

Although a definitive cutoff for what is to be included in the edible environment cannot be provided, it is arguable that it is valuable for at least four reasons, no matter how we come to individuate it from time to time. But, what is it really that is of

---

[7] To be clear, I am not suggesting that the aesthetic appreciation of nature necessarily implies the recognition of an the intrinsic value of some natural elements, nor that all works of art illustrating nature do illustrate nature's intrinsic value; at a minimum, since the biodiversity of the edible environment depends on its relationship to human tending, showcasing and valuing it, *per se*, is also a mean to showcase and value *per se* human efforts to establish a meaningful relationship with the edible landscape; I am more modestly claiming that certain works of art can serve as illustrations of the view that nature has an intrinsic value.

value? And, are there any theoretical or practical conflicts in the items that we are seeking to foster? We shall address these two questions, in order.

With respect to the first question, we shall distinguish three kinds of items that are typically regarded as valuable in discussion of the edible environment. (1) The first, more traditionally valued kind of item is the variety or breed as established by means of reproduction. It is under this regard, for instance, that we shall include the conservation of the thousands of breeds of animals reared by humans over the course of millennia; ditto for the thousands of varieties of beans, potatoes, tomatoes, corn, and other plants; the hundreds of mushrooms that humans consume; the thousands of varieties of fungi, yeasts, and strains of bacteria coopted for food production. The problem with this proposal is that it is often controversial whether some characters of a plant or an animal are novel to the point of constituting the foundation of a new breed or a new variety.[8] The issue had been touched upon also by Darwin in the *Origins of Species*, especially Chapter II. Is a variety a cluster of organisms that has the potentiality to become a new species in a near future? That seems doubtable in the case of most edible organisms. Are varieties distinguishable at the genetic, phenotypic, behavioral, ecologic, nutritional, gustatory level? Should we pick varieties based on their significance for a certain culinary history, for their relationships with surrounding ecosystems, or rather for more arguably intrinsic characteristics of the product? Notice, finally, that to intensify the efforts to preserve a variety can imply to weaken it, because it may make it increasingly dependent from humans.

(2) In recent years a new method for marking the diversity of an item within the edible realm has come to be employed: it traces the genetic specificity of a variety of plants, of the breed of an animal, or of a microorganism. Thus, a clone of – say – Sangiovese grapes can now be identified not in terms of its phenotypic traits and ancestral history, but in terms of certain genetic traits that arguably are responsible for its characteristic phenotypic traits, such as the size of its fruits, its color, its skin, or a certain gustatory quality (cfr. also Borghini 2014 on this point). Although this method of identifying an item may seem similar to the one based on breeds and varieties, it is actually quite different. Indeed, breeds and varieties are essentially linked to ancestral history; on the other hand, fixing the identity on the basis of a selected number of genetic features is compatible with cis-genesis, cloning, and other potential forms of bioengineering. Hence, the identity of a certain breed of cattle would be fixed in terms of its genetic characteristics, no matter how the cattle would come into existence (actually, no matter whether the cattle ever came into existence or whether, instead, some of its cells where cultivated in a lab; on lab-grown meat, see Van Mensvoort and Grievink 2014).

(3) A third and last kind of item that may be worth fostering in order to foster the biodiversity of the edible environment are procedures and techniques for breeding and tending plants, animals, and microorganisms. Hence, the different manners by

---

[8] The more technical definition of 'cultivar' provided for plants in ICNCP (2009, Chapter II Art. 2.3) does not help here, because it still relies on a judgment regarding the novelty of the plant character.

means of which humans have facilitated, reared, and coopted new breeds of animals, new varieties of plants, and clusters of microorganisms. Should techniques employed within bioengineering be included in this list, too? Should they receive equal weight with respect to older methods?

To the purposes of the present discussion, which aims at framing a philosophical discussion of biodiversity when it comes to the edible environment, it is important to point out that there are some incompatibilities among the three kinds of items that may be targeted for being fostered. I have hinted at one incompatibility already when presenting genetic specificity. If the policy of an institution is to foster the continuation of existence of certain genetic traits, that may imply to have to change procedures and techniques for tending it, as well as changing its reproductive history (hence, what are commonly regarded as breeds or varieties); for example, some speculated that in order to keep producing Champagne in Champagne, farmers will have to introduce genetically modified clones of grapes, possibly employing different techniques for planting (and perhaps harvesting and processing) them. On the other hand, concentrating on certain methods of, say, wine production, will typically imply that at some moment farmers will have to discard clones that are not in sinking with relevant changes within the ecosystem of production, thereby also compromising the genetic identity of the clones. Finally, focusing on breeds and varieties based on ancestry, implies embracing genetic changes over time as well as methods of production that would best meet such changes. The upshot of this analysis is that, when issuing policies for fostering the biodiversity of (some part of) the edible environment, it is relevant to specify both which kind of items are to be fostered and to what extent the kinds of items that are not to be fostered should be kept into account into the measurement and intervention efforts. This is far from being accomplished by the extent literature on the topic as well as by extant policies, such as the Convention on Biological Diversity and the International Treaty on Plant Genetic Resources for Food and Agriculture.

20.6   Conclusions

We shall at last return to our initial three questions and suggest answers based on the considerations made thus far. *(1) Are there living entities that should be excluded from measurements of biodiversity as well as from conservation efforts?* When it comes to the diversity of the edible environment, the first suggestion is to consider the importance of so-called wild species, which to date play a critical role in integrating agricultural and industrial produce in most societies, constituting also an important back-up safety net for food security purposes. A second suggestion is to proceed with great care when it comes to drawing cutoff points between items that are natural enough to deserve inclusion in an inventory of food biodiversity and items that are not so; it seems most prudent to proceed by devising cutoff points that are suitable to specific sub-domains; these can be individuated on different grounds, such as biological taxa (e.g. cucurbitaceae, beans, mushrooms) or methods of production (e.g. grafting, sexual reproduction, genetic modification). Thus, we have multiple possible inventories to choose from, giving rise to our second question.

(2) Should the criteria for inclusion in a measurement or intervention be context-dependent or context-independent? For instance, could different criteria be selected depending on circumstances? A successful discussion of the matter, I submit, would demarcate as clearly as possible what are the conceptual and axiological differences between the criteria, as well as their potential practical consequences. It is important to remark here that the diversity of the edible environment is deeply entrenched with human cultures, so that the criteria for biodiversity measurement must reflect human perspectives within different societies, embedded in the conceptions of plants, animals, and dieting.

(3) What are the grounds for preferring certain (possibly more familiar) forms of life over others? For instance, do such preferences rest on efficiency, or perhaps on some emotional or spiritual connection? This question addresses the values that are involved across possibly different context of evaluation, e.g., food sovereignty, food security, and gastronomic pleasure. It is important to explore how such values differ across societies and whether convergence over a few selected values is a desirable goal, or if lack of convergence is actually more fruitful for the purposed of the biodiversity of the edible environment.

The new agricultural technologies introduced by the Green Revolution between the 1930s and the 1960s, followed by the more recent innovations in biotechnology, along with an increased capacity of transportation, have deeply subverted the relationship between humans and food. Such a drastic rupture is forcing a rethinking of that relationship, and a careful consideration of which items we shall conserve and why. This essay aimed at introducing a philosophical a frame for assessing the biodiversity of the edible environment, and pointing at a number of questions that seem in need of being addressed in the near future.

References

Alkon, A. H., & Mares, T. M. (2012). Food sovereignty in US food movements: Radical visions and neoliberal constraints. *Agriculture and Human Values, 29*, 347–359.

Berry, W. (1990). The pleasures of eating. In *What are people for* (pp. 145–152). New York: North Point Press.

Bharucha, Z., & Pretty, J. (2010). The roles and values of wild foods in agricultural systems. *Philosophical Transactions of the Royal Society B, 365*(1554), 2913–2926.

Borghini, A. (2014). Geographical indications, food, and culture. In P. B. Thompon & D. M. Kaplan (Eds.), *Encyclopedia of food and agricultural ethics* (pp. 1115–1120). New York: Springer.

Borghini, A. (2017). Hunger. In P. B. Thompon & D. M. Kaplan (Eds.), *Encyclopedia of food and agricultural ethics* (2nd ed., pp. 1–9). New York: Springer. (ONLINE FIRST).

Brussaard, L., de Ruiter, P. C., & Brown, G. G. (2007). Soil biodiversity for agricultural sustainability. *Agriculture, Ecosystems and Environment, 121*, 233–244.

Chappell, M. J., & La Valle, L. A. (2011). Food security and biodiversity: Can we have both? An agroecological analysis. *Agriculture and Human Values, 28*, 3–26.

Chemnitz, C., & Becheva, S. (2014). *The meat atlas. Facts and figures about the animals we eat.* Ahrensfelde: Möller Druck.

Conservation and Sustainable Use of Agricultural Biodiversity. A Sourcebook (2003), CIP-UPWARD – International Potato Center: Laguna.

Costanza, R., et al. (1997). The value of the world's ecosystem services and natural capital. *Nature, 387*, 253–260.

Diamond, J. (2002). Evolution, consequences and future of plant and animal domestication. *Nature, 418*, 700–707.

Fromartz, S. (2006). *Organic Inc. Natural food and how they grew.* Orlando: Harcourt Books.

Grivetti, L. E., & Ogle, B. M. (2000). Value of traditional foods in meeting macro- and micronutrient needs: The wild plant connection. *Nutrition Research Reviews, 13*, 31–46.

ICNCP. (2009). *International Code of Nomenclature for Cultivated Plants* (8th ed.). Vienna: International Society for Horticultural Science.

Jarosz, L. (2014). Comparing food security and food sovereignty discourses. *Dialogues in Human Geography, 4*, 168–181.

Kowalsky, N. (Ed.). (2010). *Hunting: In Search of the Wildlife.* Oxford: Wiley-Blackwell.

Ludlow, C. L., et al. (2016). Independent origins of yeasts associated with coffee and cacao fermentation. *Current Biology, 26*, 1–7.

Mazhar, F., Buckles, D., Satheesh, P. V., & Akhter, F. (2007). *Food sovereignty and uncultivated biodiversity in South Asia.* New Delhi: Academic Foundation.

McNeely, J. A., & Sherr, S. J. (2002). *Ecoagriculture: Strategies to feed the world and save wild biodiversity.* London: Island Press.

Millenium Ecosystem Assessment (MEA). (2005). *Current state and trends.* Washington, DC: Island Press.

Petrini, C. (2013). *Slow food nation: Why our food should be good, clean, and fair.* New York: Rizzoli Ex Libris.

Pollan, M. (2001). *The botany of desire. A plant's eye view of the world.* New York: Random House.

Pollan, M. (2006). *The omnivore's dilemma. A natural history of four meals.* New York: Penguin Press.

Siipi, E. (2016). Unnatural kinds: Biodiversity and human modified entities. In J. Garson, A. Plutynski, & S. Sarkar (Eds.), *The Routledge handbook of philosophy of biodiversity* (pp. 125–138). New York: Routledge.

Sperber, D. (2007). In E. Margolis & S. Laurence (Eds.), *Seedless grapes: Nature and culture* (pp. 124–137). Oxford: Oxford University Press.

Stiegler, B. (2006). Take care (Prendre soin). (trans. by S. Arnold, P. Crogan and D. Ross), *Ars Industrialis*: http://arsindustrialis.org/node/2925

Thompson, P. B. (2015). *From field to fork: Food ethics for everyone.* Oxford: Oxford University Press.

Van Mensvoort, K., & Grievink, H. J. (2014). *The in vitro meat cookbook.* Amsterdam: BIS Publisher.

Vitalini, S., Tome, F., & Fico, G. (2009). Traditional uses of medicinal plants in Valvestino (Italy). *Journal of Ethnopharmacology, 121*, 106–116.

Wrangham, R. (2009). *Catching fire: How cooking made US humans.* London: Profile Books.

21

Biodiversity and Sovereign States

Markku Oksanen and Timo Vuorisalo

Abstract Many key concepts in conservation biology such as 'endangered species' and 'natural' or 'historic range' are universalistic, nation-blind and do not implicate the existence of geopolitical borders and sovereign states. However, it is impossible to consider biodiversity conservation without any reference to sovereign states. Consequently, the units of biodiversity and their ranges transform into legal concepts and categories. This paper explores the area that results from this transformation of the universalist idea into national policy targets. Conservation sovereignty denotes to right of each state to design and carry out its own conservation policies. To illustrate the problematic nature of conservation sovereignty, the paper focuses on two cases where the borders and the state play the key role: (1) the global division of conservation labour and (2) assisted migration. All in all, this paper takes a critical look upon the anomalies in universalism and conservation sovereignty.

Keywords Species · Environmental ethics · Natural resources · Assisted migration · States

M. Oksanen (✉)
Department of Social Sciences, University of Eastern Finland, Kuopio, Finland
e-mail: markku.oksanen@uef.fi

T. Vuorisalo
Department of Biology, University of Turku, Turku, Finland
e-mail: timo.vuorisalo@utu.fi

21.1   Introduction

Many commonly used concepts in conservation biology such as 'endangered species' and 'natural' or 'historic' range[1] are universalistic, nation-blind and do not implicate the existence of any geopolitical borders or sovereign states.[2] To ignore the current nation-state system and to consider conservation of biodiversity without any reference to states would, however, be unsatisfactory. States are self-determining actors and the principal possessors of biological resources in their territories. At the international level, sovereignty is manifested in the international treaties and declarations approved by the states. By these treaties and declarations, states commit themselves to certain responsibilities and thus voluntarily restrict the ways of acting open to them. At the national level, sovereign states implement these agreements within their jurisdictions, that is, within their established geopolitical borders. From this constellation, a vital point emerges with respect to biodiversity conservation: the units of biodiversity and concepts ascribing their ranges transform into legal concepts and categories that inform policies and practices. This perspective regards sovereign states as the only relevant legal actors. The transformation thus occurs within, and is organised by, the sovereign states.

In creating national policies for biodiversity conservation, sovereign states act either alone or in close collaboration with other states (consider the EU). A global division of conservation labour arises out of joint multiple actions by states.[3] The fundamental idea is that each country, as a sovereign actor, is in charge of the biodiversity within its territory while the biodiversity outside the territories of sovereign states (the high seas, the Antarctica) as well as migrating biodiversity (waterfowl, whales) are subject to transnational decision-making, if any.

In this chapter, the traditional thinking will be modestly challenged in two ways. On the one hand, we argue that the global division of conservation labour in its present form is not always efficient from the conservation perspective if each country only focuses on safeguarding its territorial biodiversity. On the other hand, we ask whether climate change (in the global perspective) could challenge the current conservation policies by requiring actions that would make state borders more porous, and applied policies more interventionistic than what they are today. We contend that in some cases successful conservation may require international translocation measures for the establishment of new populations outside the historical ranges, and geopolitical territory, of particular species.

---

[1] There are a plenty of other attributes to describe ranges such as 'indigenous' and 'native', some of which may be more sensitive to current geopolitical structure than the notions of natural and historic (on their differences, see Siipi and Ahteensuu 2016).

[2] Smith's (2016) analysis manifests a universalistic viewpoint concerning the ethics of endangered species preservation.

[3] In addition to this expression being powerful in its own right, it articulates and explicitly includes human-dependent form of biodiversity. In most cases, this biodiversity literally results from human labour.

The aim of the chapter is to explore issues that result from the transformation of the universalistic idea into national policy targets the foci of which are not merely species universally understood but a wider variety of different "conservables". To understand what these conservables are, we come across the political dimensions of biodiversity conservation. In the first section, we discuss the idea of state sovereignty and its relation to the control of natural resources and biodiversity. The second section, in turn, presents a typology of sovereignty in the context of biodiversity conservation. In the third section, we examine the global division of conservation labour and its insensitivity for the issue of prioritisation, and the resulting obvious need to transform conventional conservation. The fourth section analyses assisted migration, or whether it is acceptable to translocate species (across the state borders) with the intention of helping them to survive global warming. A short conclusion ends the discussion.

Four clarifying remarks on the nature and scope of our inquiry need to be made. First, our approach is multidisciplinary and focuses on conceptual and theoretical problems arising from sovereignty in the context of biodiversity conservation. We also examine some real-life examples. Second, the transformation from scientific descriptions to legal categories and to conservation success may seem simple but is in reality a complicated and twisted issue because corruption in land-use decisions is widespread and it is difficult to prevent poaching and illegal wildlife trade. Although tackling illegalities is undoubtedly relevant to policy design, it is outside our main analysis. Third, our approach is stated-centred and thus extremely constricted. For a more comprehensive picture, the nonstate or civil society actors such as citizens, academics, non-governmental organisations, state-funded think tanks like the OECD (the Organisation for Economic Co-operation and Development) and transnational companies should be taken into account. Fourth, issues of security and safety, in particular the border control of the import of unwanted or hazardous biomaterial, have been and are important components of sovereignty; they are outside our scope of analysis. Therefore, keeping these remarks in mind, the picture we paint of sovereignty is at best sketchy and filled with promises that may never actualise; it is, nonetheless, a useful starting point for further analysis.

21.2   Biodiversity in the World of Sovereign States

Sovereignty over natural resources within the state territory is today an established principle in international law. The concept of 'sovereignty' dates back to the late sixteenth century and to the French political theorist Jean Bodin who famously wrote that, "sovereignty is the most high, absolute, and perpetual power over the citizens and subjects in a Commonwealth" (cited in Turchetti 2015). In actual politics, sovereignty became a leading principle in international law as a result of the Westphalian peace in 1648; hence the international system of sovereign nation-states is still known as the Westphalian system. In the historical context of Bodin and other peace negotiators, the unchallenged presumption was that absolute monarchy is a legitimate form of government. This aspect is not relevant to our analysis despite the facts that many biodiversity-rich countries lean towards absolutism and their democracy, civil societies and status of minorities can be questioned and the global developers' and resource buyers' voices are often compelling. In modern use, sovereignty is typically understood as a form of power that belongs to the state indivisibly and above other powers. In this sense, sovereignty expresses the idea of the right to self-determination that is hold by the nation-state over territory, natural resources and the peoples who inhabit the area. The sovereignty of the nation-states also guarantees a legal personhood for this entity in the international legal system, that is, it is externally independent and can exercise power within a community (Endicott 2010). Because of sovereignty, states are in the position to enter voluntarily into binding, action-limiting and, in some cases, external interference entitling conventions (Shue 2014, 146).

A key issue in discussions on sovereignty has been the control of natural resources. Natural resources are thought of as instrumental for the full exercise of self-determination: hence, without possibility to exclude other states (and nonstate trespassers) from using natural resources within their territories, states cannot be truly independent beneficiaries of their own natural wealth. This idea was particularly powerful in the post-World War II period of decolonization and the dissolution of the British, French, Japanese and other empires. In addition, resource scarcity was a matter of mounting concern, which inspired the US President Truman set up by the Materials Policy Commission in 1951. The Commission's analysis *Resources for Freedom* (1952) reflected the general pessimistic mentality with respect to resource availability now and in the future although it recommended policies supporting economic growth. (Andrews 1999, 182–83.) To consolidate the ties between national independence and self-determination and the control of natural resources, the General Assembly of United Nations adopted resolution 1803 (XVII) on the "Permanent Sovereignty over Natural Resources" in 1962.

Sustainable development became a truly global issue by the publication of Brundtland's Commission report *Our Common Future* in 1987. According to it, the current use of resources must not come about at the cost of the resource use, or welfare, of the future generations. It strongly influenced the contents of the Convention on Biological Diversity, signed in Rio de Janeiro in 1992. According to Article 2 of the Convention, biological resources include genetic resources, organisms or parts thereof, populations, or any other biotic component of ecosystems with actual or potential use or value for humanity. Furthermore, sustainable use of these resources "does not lead to the long-term decline of biological diversity, thereby maintaining its potential to meet the needs and aspirations of present and future generations". Later, a resource-based approach to biodiversity conservation has been very strong in the Ecosystem Approach that is a framework for action under the Convention.

The question is then: in what sense is biodiversity a natural resource? It seems straightforward to reason that if the concept of natural resources covers all resources that are biological, and if the concept of biological resources, in turn, includes biodiversity in all of its manifestations, then biodiversity is a natural resource. This view is emphasised by the Ecosystem Approach that focuses on the importance of ecosystem services that in fact cover all major biological processes in their natural environments. Not all resources are tangible; the category of cultural services, as a component of ecosystem services, includes historical, spiritual, educational and recreational values that ecosystems have but which can be damaged through the loss of biodiversity. Obviously, the convertibility of such cultural values into resources, or monetary values for that matter, is problematic, perhaps with the exception of eco-tourism or popular historical monuments that clearly have a market value. Many authors, however, resist this way of considering biodiversity merely as a resource (see e.g. Wood 1997) and the associated rather explicit anthropocentric attitude to the rest of nature.

When we adopt the conception of state sovereignty – a conception that is at least historically anthropocentric since it entitles 'peoples and nations' to utilize their natural resources – it depends on states what meanings they attribute to biodiversity in practice. This framework, however, emphasizes for the above mentioned historical reasons the status of biodiversity as an instance of natural resources. It is clear, however, that there are natural resources that do not fall into the category of biodiversity conveniently (e.g. water and non-renewable mineral resources) and the significance of biodiversity is not exhaustively reducible to its resource character. For this reason, when we talk about biodiversity within the framework of sovereignty, we should not consider it merely as a bundle of natural resources but having significance beyond their "resourceness", a point also made in the opening lines of the Preamble to the Rio Convention. An interesting question is which parts of biodiversity fall outside the popular concept of ecosystem services. To make these conservation dimensions more explicit, we purport the idea of conservation sovereignty.

Conservation sovereignty, as a political idea, stands for the right of each state to design and carry out its own conservation and related natural resource policies, as if there were no transnational regulation. Since there is, however, transnational regulation agreed upon by the sovereign states, though not necessarily by all of them, the question arises whether there can then be sovereignty with respect to biodiversity and its conservation. The paradox is apparent and there are rival attempts to tackle it. As Endicott (2010, 246) has noted, "state sovereignty seems both to demand the power to enter into treaties, and to rule out the binding force of treaties." It is clearly analogous with the better-know philosophical dilemma of whether the freedom of a human individual includes the possibility to enslave oneself for a fellow human, as Endicott (2010, 246) points out. We follow Endicott's (2010, 258) conclusion that sovereignty and participation in global agreements and international law are "at least potentially compatible" although the function of these agreements and laws is to give directions to domestic laws and policymaking and to guide interactions between states. As Shue (2014, 143) puts it, "sovereignty is not some mystical cloud that either envelops the state entirely or dissipates completely; there are bits and pieces of asserted sovereignty." A look at the recent history makes one think that there are no theoretical tensions: the processes of decolonization and the formation of the system of over 200 sovereign states have occurred simultaneously with the growing number of international environmental treaties (Frank 1997).

The paradoxical dimensions of sovereignty are also recognizable in the Rio Convention. According to the Preamble, "States have sovereign rights over their own biological resources" and thus it merely expresses the established principle that the biological resources in state territories are freely at disposal of the state. The previous passage, however, outlines reasons for restricting state sovereignty and the free use of these resources, since "the conservation of biological diversity is a common concern of humankind". Nevertheless, this paradox is a milestone in the development of international regulation concerning biodiversity. Given the long UN history on the issues of sovereignty and natural resources, the authors of the Preamble were fully aware of tensions between national interests and universal concerns and the essential differences between objects of human interests. The novel expression 'common concern' reflects the negotiators' worry about the state of biological diversity beyond specific geographical areas and resources to which the already established concepts of common area and common heritage apply (see Brunnée 2008).

The Preamble of the Rio Convention effectively captures the two-dimensional nature of global conservation efforts: it is international and domestic at the same time. Within the European Union, two-dimensionality is most clearly manifested in the Natura 2000 conservation area network, established by the Habitats Directive in 1992. The duality between nationalism and internationalism has its roots in the origins of modern conservation movement in the late nineteenth century.[4] Ever since the creation of the Yellowstone National Park in 1872, most countries have followed the model and selected areas to sustain wilderness and pristinity. In spite of its gradually increasing popularity in the USA (see Nash 2001, 108–21) and other countries, the national park movement was essentially a nationalistic enterprise that emphasized each country's unique nature values – in some cases compared with those of neighbour states. As Sheail (2010, 12) put it: "*National* parks presuppose sovereign nation states".

The idea ultimately reached the Old Continent with the first European national parks founded in Sweden in 1909.[5] The famous explorer A. E. Nordenskiöld had already in 1880 urged the establishment of 'state parks' in Nordic countries to preserve samples of fatherlands' pristine nature for the future generations (Palmgren 1922). The patriotic tone was unmistakable in the essay of the Finnish State Conservation Inspector, Dr. Reino Kalliola, who wrote in the first issue of *Suomen Luonto* – the journal of the Finnish Association for Nature Conservation – that, "the richness and beauty of the Finnish nature is our shared and precious heritage that everyone of us is obliged to cherish" (Kalliola 1941, 20; also Kalliola 1942). Although similar nationalistic tones were probably heard in conservationist circles across the world in the nineteenth century, also the first important multilateral

---

[4] In this analysis, we try not to identify the origin of conservation practices and we leave out the discussion on imperialist roots of early conservation (see Grove 1995).

[5] A somewhat parallel development took place in Britain, with the Establishment of the National Trust in 1895. Although emphasis of the National Trust has been on preservation of cultural heritage, also areas of natural beauty have been preserved (Sheail 2010).

conservation agreements, such as for instance the Paris Convention for the Protection of Birds Useful to Agriculture (1902), date back to that period (Lyster 1994).

Some early pioneers of conservation movement were active both internationally and nationally. The protection of migratory birds is a case in point. Even before the independence of Finland in 1917, the leading Finnish conservation pioneers had close relations to colleagues abroad and in different occasions pursued internationally defined objectives at the national level. Dr. Johan Axel Palmén, Professor of Zoology in Helsinki, took great interest in the 1st International Ornithological Congress in Vienna in 1884. It is notable that the delegates of the conference attended as individual citizens, as respected members of the scientific community and not as official delegates sent by their respective governments. The governmental acceptance of conservation matters was, however crucial and official participation increased gradually. It is illustrative that The International Council for Bird Preservation (ICBP; from 1993 on, Birdlife International) was founded at the Finance Minister's home in London in 1922 (Birdlife 2017). Accordingly, the idea of national representation in international meetings was stronger providing a better basis for national action on bird conservation. To return to Palmén, the year following the Vienna conference, he published a seminal paper that outlined a plan, based on the conference proceedings, for a reliable collection of nationwide data on bird species distribution and abundance in all regions of the country (Palmén 1885). Palmén's programme turned out to be very successful (Vuorisalo et al. 2015). Later, Palmén (1905) proposed setting up a national conservation society (this happened in 1938), and protecting the endemic Saimaa ringed seal population (legal protection 1955, see Case 1). After independence in 1917, it seems that the attention of Finnish conservationists turned almost entirely to domestic affairs, with a strong emphasis on the establishing and expanding of the national and nature park network (Vuorisalo and Laihonen 2000).

Scientific communities of specific disciplines are universal and, in principle, independent of governments. However, without governmental support their goals, both scientific and non-scientific, are difficult to reach. Likewise, as compared to the powers at the disposal of the state, the international community is rather weak in environmental matters. One reason for this is structural and institutional: there is no global government with the right to tax persons or states or penalise those parties who violate the global rules. The possibilities of ruling sanctions are limited. The ambition to reach unanimity in policy-making often leads to vague compromises, and when unanimity is not aimed at, the risk of free-riding (benefiting without taking responsibility) and gaps in policies is apparent. As Simon Lyster states, "the international community ... has no legislature capable of formulating laws binding on individual States or their peoples without their individual consent" (1994, 3). What is the ensuing nature of conservation sovereignty in such a situation? The answer is that there have been and still are rival conceptions very vivid in the political debates.

21.3 Three types of National Sovereignty in Biodiversity Conservation

The starting point of sovereignty with respect to biodiversity is that biodiversity constitutes an instance of natural resources. Of course, in the background is the policy of priority setting based on the conservation value of biological units, that is, of subspecies, species and biotypes. Within biology, the definitions of biodiversity and its units have been debated continuously since the 1980s, as the existence of this volume also indicates (see also, Gaston and Spicer 2004). Whatever the units, we may call them here conservables. As pointed out earlier, biodiversity is not merely a resource but also a conservable. The most crucial distinctive factor between these two concepts is that conservables have such significance for humans that is not entirely reducible to crudely instrumental or purely monetary values, whereas the notion of resource specifically implies both of those values. In the context of modern market economy, resources are resources to someone whose access to the resources depends on established property and market relations. Although conservables can also be classified as resources, their status and significance is not limited to their 'resourceness'; consider as an example cultural landscapes with exceptional diversity (cf. Oksanen and Kumpula 2018). Thus, the adopted approach should be wide enough so as to include conservation policies that take into account these non-resource dimensions. Conservation sovereignty, distinctively, refers to the right of each state to design and carry out its own conservation and natural resource policies. One such option, within a strong conception of sovereignty, is that the state decides not to have any conservation policies and gives free hands for the user of natural resources as long as inflicted harms are at a tolerable level. In today's world, such an option would stand out as exceptional.

To make precise the contrasting understandings about sovereignty and conservation, we distinguish between three kinds of sovereignty. These types are both historical and theoretical constructions. One can also envision, as many have done, global systems without putting states in the central positions and having some kind of a world government; such a system would undermine the talk over sovereignty as we know it and is therefore not analysed here.

Traditional conservation sovereignty ('brute nationalism') refers to the traditional system, stemming from the nineteenth century, where each country creates its conservation legislation and network independently of other countries. The pioneering phase of national park movement across the world clearly represents this category. In each country, national parks were established based on the country's own legislative system. Decision-making was thus strictly national and any country having no interest in adopting conservationist policies was at liberty to do so.[6] The aim

---

[6] Henry Shue, in discussion on climate changes policies, is critical of sovereignty that allows states to pursue economic growth, if they choose to. He writes that "there ought to be external limits on the means by which domestic economic ends may be pursued by states, limits that ought to become binding on individual sovereigns irrespective of whether those sovereigns wish to acknowledge

of self-sufficiency naturally does not exclude possibilities that some influences spread from one country to another. Moreover, cooperation between states is reasonable since some activities can generate transboundary harms and many resources (migratory species, boundary rivers, for instance) are multi-territorial. In those cases, bi- and multilateral resolutions may be agreed upon. In social studies on conservation and natural resource use, the classical research themes include the analysis of the conflictual relationships between the central power and the localities and what kind of institutional arrangements would work best in given conditions. The traditional conservation sovereignty can be understood to imply a strongly state-led approach to conservation in which local-level interests and arrangements, including those of the indigenous peoples, become overridden. On the other hand, often, but not always, localities are the best managers of extant biological diversity and decisions from afar can lack adequate local acceptance. In traditional conservation sovereignty, it is a domestic issue how these challenges are met (although there can be other relevant restrictions based on international law such as human rights).

The traditional conservation sovereignty is deficient because of the biospheric nature of biodiversity and its components. As mentioned earlier, historically international practices that aimed at bird conservation were developed very early. There were also debates about the inexhaustibility of other migratory and often highly exploited species and, respectively, a need for international regulation in hunting, fishing and whaling (Lyster 1994). What this has brought about is *internationally regulated conservation sovereignty* ('externally constrained nationalism'). According to it, countries voluntarily participate in international conservation agreements and pursue the harmonisation and unification of conservation efforts at the regional and global levels. This is the system characterized by most of today's states' conservation policies (cf. Lyster 1994). For instance, the Convention on Biological Diversity has now (as of June 2017) 196 Parties that have ratified the treaty.[7] Internationally regulated conservation sovereignty has prevailed ever since the Stockholm Conference of 1972 that launched unprecedented international environmental activity. Although the principle 21 of the Stockholm Declaration declares that, "States have … the sovereign right to exploit their own resources pursuant to their own environmental policies", the same principle continues by requiring that developmental activities do not damage the environment. Many international environmental treaties acknowledge broad principles that guide the construction and implementation of more specific norms. These principles include the recognition of the duties to future generations, the prevention of environmental harms, the polluter-pays principle, cooperation among states and ideas about burden sharing. More recently, the development of international environmental law has focused on establishing institutions and procedures through which scientific communities and new research results can be better accommodated into policies. The flagship model is the IPCC, the name of which refers to collaboration between sovereign states – Intergovernmental Panel on

---

them, just as sovereigns are already bound by both legal and moral rights against domestic use of torture […]" (Shue 2014, 150).

[7] See the list of signatories here: https://www.cbd.int/information/parties.shtml

Climate Change. The model was adopted to biodiversity conservation when the Intergovernmental Science-Policy Platform on Biodiversity and Ecosystem Services (IPBES) set off in 2012. Currently (as of June 2017), IPBES has 126 states as its members.

The system where a state has only a partial sovereignty over its natural resources can be called *federal conservation sovereignty* ('regionally constrained nationalism'). In this system, states share a major portion of their conservation legislation and the compliance with supranational laws is monitored and sanctioned. The European Union is the prime example of this case. According to article 47 of the Treaty on European Union (the Treaty of Lisbon, 2007), the Union recognises itself as a legal person with rights to join international conventions, for instance. However, to state that it is a sovereign state in its own right is a contentious federalist statement and seen to contradict the sovereignty of the member states. Therefore, there is no such official statement. Since it is not our main topic to tackle this sensitive issue and suggest appropriate political moniker, it is a safe bet to characterize it as a closely-knit alliance of sovereign states with sovereignty in selected international issues and with power to circumscribe national sovereignty over agreed areas of public policy (cf. Philpott 2016). At the Union level, the principal issues of biodiversity are being dictated through 'directives'. The idea of the directive is that the addressed member states must adopt into their legislation the designated goals while the choice of form and methods of achieving them belongs to national authorities. The Birds and Habitats directives are the main legislative tools for biodiversity conservation in the EU, and in addition to habitats, their focus is on species, as the official website summarises: "The Habitats Directive ensures the conservation of a wide range of rare, threatened or endemic animal and plant species" (European Commission 2017). Many federal states are legal persons in international law whereas their provincial components are not. In the Westphalian system, these actors are not sovereign and are therefore excluded from foreign politics. However, one of the elements of globalisation is the increasing cooperation between cities and regional actors across national borders and in some cases in explicit opposition to the decisions of the central government. There are numerous comparative studies on the EU and existing federal states like the USA on specific policy areas. It is easy to parallel, for instance, the Birds and Habitats Directives with the Endangered Species Act of the USA: both are regulations from the central government. Such a parallelism can, however, be a simplification. With respect to biodiversity, in the United States an individual state and municipalities may adopt rather independent policies; whereas in the European Union the EU decrees and directives strictly control what a member country can rule in its national legislation (cf. Wells et al. 2010).

As these three contrasting views on sovereignty indicate, the development of supranational and international environmental law constrains the opportunities to enforce policies solely on the national basis. The pure or brute form of sovereignty has become, as has been noted from time and again, an obsolete idea as soon as the ecological ideas have matured enough. In international studies, discussion on states and their standing has been enduring. Though sovereignty is a kind of trump card, the international processes and institutions of governance have evolved to tackle the complex problems of biodiversity loss. Nevertheless, sovereignty should not hide from us the complexities of vocabularies, institutions and practices in international biodiversity management, and from its somewhat decentralised character (see e.g. Ostrom 1998).

21.4   Case 1: Global Division of Conservation Labour: The Prioritisation Problem

Interestingly, the case of biodiversity has some structural commonalities and substantial convergences with the idea of human rights. Consider Beitz's formulation of what he calls the two-level model of human rights: "The two levels express a division of labour between states as the bearers of the primary responsibilities to respect and protect human rights and the international community and those acting as its agents as the guarantors of these responsibilities." (Beitz 2009, 108.) To some extent, this has been apparent also in the field of international environmental law (Lyster 1994). As applied to biodiversity conservation, such a division of labour could mean that states bear primary responsibility for biodiversity conservation within their territory while the international community may set general guiding principles for conservation efforts in multilateral agreements and acts as a guarantor of this responsibility. As a result, some division of labour in biodiversity conservation develops between sovereign states and the international community.

In conservation policy, the idea of the global division of conservation labour refers to the emergent properties of conservation and how they are manifested through adopted collaborative and domestic practices for instance in the ratification processes of multilateral environmental treaties. Fundamentally, each state is a sovereign state with rights and obligations to accomplish within its territory. On the one hand, sovereign states have rights to resources; on the other hand, and in our analysis more importantly, each nation-state is responsible for protecting the biodiversity within its borders. We can take this literally and thus have a rather mechanistic approach to biodiversity conservation. This means that the conservation value of policy targets, or conservables, is defined nationally based on their abundance and distribution within the state borders.

Reflecting the general tone of this edited volume, we reckon that the emphasis in policymaking has traditionally been on species although there are more nuances to it. As the main goal of conservation efforts is to conserve evolutionary potential, we often need to be concerned about possible management units below the species level. Such units have been called Evolutionarily Significant Units (or ESUs), and may be defined as partially genetically differentiated populations that are thought to require management as separate units (Frankham et al. 2002). Biologically, it may be a matter of taste whether such units are called species, subspecies, or simply local populations. However, terminology matters in conservation policy. It may be easier to get support for conservation of a separate endemic species (that may even become a national symbol) than for an obscure local population. Under such circumstances, 'species as targets of conservation policies' may be created through campaigning, policies and practices, not purely scientifically. A case in point is the Saimaa ringed seal (*Pusa hispida saimensis*) in Finland, first scientifically described in the late nineteenth century (Nordqvist 1899); it is either a "critically endangered" species or a rare fresh-water population (or subspecies) of the "least concern" ringed seal. Today, the Saimaa ringed seal is a symbol of national conservation efforts in Finland even without being a species proper. Because we do not want to deny the significance of its conservation, our point is the following: if the populations and subspecies are classified as species proper, this is not necessarily a scientific error but rather an inaccuracy based on inherent ambiguity of taxonomic classifications. As this example indicates, biodiversity is a political concept that relates to existing political systems in a way that may affect the scientific basis of conservation.[8]

Another type of conservation controversy arises when the population of a certain species is endangered locally or regionally, but not globally. Consider the following example of species preservation where the targeted species occurs across the Eurasian taiga but is rare within the European Union. In the EU, the Siberian flying squirrel (*Pteromys volans*) only occurs in Finland and the Baltic states (Estonia and Latvia). Despite its universal Red List status as 'least concern', the mechanistic application of global division of conservation labour calls for its prioritisation in national policy. In Finland, the flying squirrel has become a symbol of public conservation battles that has caused trouble to, in particular, building and road construction (Hurme et al. 2007). The big question now is: does it really make sense to mechanistically follow the division of conservation labour between sovereign states, especially in situations like the aforementioned?

In the EU, the Habitats Directive defines as an overall objective of conservation measures the maintenance or restoration of natural habitats and populations of Community interest at a favourable conservation status (Mehtälä and Vuorisalo 2007; Epstein et al. 2016). This objective is achieved through a division of labour between member states which, in the case of the Siberian flying squirrel, means that the above-mentioned three states are responsible for maintaining the conservation status of the species within the Union at a favourable level. Again, in this case of federal conservation sovereignty the target is set only taking into account the species' status within the Union, with no regard of its thriving main population in the Russian Federation.

There seem to be three basic arguments in conflict here: efficiency, lack of means of global prioritization of conservation targets, and risk of erosion of the division of conservation labour. Efficiency here points to the chronic resource scarcity in conservation and the following necessity to make prioritisation decisions from a universalistic perspective and by ignoring national borders. However, although there is no

---

[8] Smith (2016) is an example of an approach focusing on endangered species so heavily that, he alleges, "sub-species are not real" (p. 4) and their identification is arbitrary. By implication, the reason for their conservation must be different from the reasons used for justifying species conservation.

lack of global and science-informed attempts for prioritization (cf. Norman Myers' 36 global biodiversity hotspots or the IUCN Red List of Threatened Species), international law does not provide any effective tools for global-level prioritization of conservation targets. Accordingly, the populations of common species in fringe areas deserve less attention. Whereas the former is crucial, the latter might affect the motivation of conservation in a negative manner. It is obvious that without any other agreement that would define the specific responsibilities, the possibility that the species is neglected emerges. Thus, these specific responsibilities must be agreed upon by all relevant parties and made explicit to avoid the vicious circle that could, at worst, lead to its extinction. A case in point of the risk of erosion of division of global conservation could be the recent policy conflicts over the Great Cormorant conservation status between the EU and some of its member states (Rusanen et al. 2011).

Obviously, from the conservation biology perspective decisions concerning the conservation of biodiversity should be made as if there were no state borders. Even the currently prevailing internationally regulated conservation sovereignty can be considered wasteful as resources are invested (sometimes massively) on local conservation efforts that have little value from the global perspective. For instance, since the 1980s lots of resources have been invested in the protection of the local White-backed woodpecker (*Dendrocopos leucotos*) in Finland, although the species continues to be common in the neighbour states of Norway (700–900 pairs) and Estonia (500–1000 pairs) (Väisänen et al. 1998; Laine 2015). Luckily, the national conservation efforts appear to have been effective, since the breeding population of the White-backed woodpeckers in Finland has clearly increased since 2010 (Laine 2015).

So it is obvious that rigid, non-adjustable nationalism has its shortcomings in today's globalized world. Moreover, we argue that under the global biodiversity crisis conservation sovereignty is becoming problematic also for two biological reasons. First, conflicting conservation priorities between countries and between the international and national level make rational (in the conservation biology sense) resource allocation very difficult. Second, the conservation area networks established by sovereign states are rapidly losing their original natural values due to climate change. The biodiversity crisis calls for an unprejudiced re-evaluation of alien species policies and assisted migration attempts that can result in some minor changes to current legislation (see Trouwborst 2014 on the EU legislation).

21.5   Case 2: Assisted Migration of Plants and Animals

Let us turn to the second issue challenging the mechanistic understanding of conservation sovereignty: the designed relocation of *alien* organisms across state borders. Considering the political restlessness caused by refugees from armed conflict areas in the Near East and the number of immigrants, a letter titled "Britain should welcome climate refugee species" appears extremely provocative. It was published in

The New Scientist in 2011, well before the Brexit referendum of 2016. The author, British biologist Chris Thomas, condensed his message in two sentences: "Some places are ideal havens for species threatened by climate change. One is Britain, and it should throw open its doors." (Thomas 2011a, b).

Thomas took sides in the recently burgeoned discussion about a new approach to biodiversity conservation: assisted migration. Assisted migration is just one of the many monikers of this particular approach; assisted colonization, managed translocation and managed relocation are among others (Hällfors et al. 2014). Indeterminable numbers of species in many countries have already begun to adapt to climate change by expanding their ranges upslope or to higher latitudes (Parmesan and Yohe 2003). This survival strategy is, however, not available to each and every species. Assisted migration roughly means that humans are to take an active role in translocating species that are believed to be at the risk of disappearance in their current range of distribution because of the impacts of global warming. The potential recipient areas are those where these species can be predicted to survive and reproduce in the future warmer climate, provided that there are no dispersal barriers or lack of time (Hällfors et al. 2014). It requires, of course, a lot of work to identify to suitable species for relocation (see Hällfors et al. 2016). Moreover, since the climate change scenarios are numerous and controversial, so are the potential recipient areas, too.

Assisted migration departs from conventional conservation policies in three ways. Firstly, unlike the established *in situ* conservation strategy that seeks to protect species within their current ranges, as the vital elements of their present or historic habitats, assisted migration is interventionist in essence. Secondly, the international legislation regarding wildlife, such as the CITES treaty and, to some extent, the Rio Convention on Biodiversity, restricts the transfer of species and/or biological material across national borders. Assisted migration, or some aspects of it, could be in conflict with current legislation although less so, if the translocation takes place within one country. And thirdly, non-native animals or plants are typically thought of as unwelcome invaders, as aliens. The national border is the most important border, although invasion can occur also within the nation-state. As Thomas' use of words exemplified, the notion of non-nativity is often constructed in terms of nationality and the role of national borders plays a greater role than the biological ideas of indigenous or historic ranges. Of course, borrowing concepts from political discourses affects how the activity will be perceived by the public.

It seems to us, thus, that we can conceptualize animal and plant species either as climate refugees or as exotic or alien invasive species. This conceptual divide seems to capture the conflicting attitudes to the ideas of plant (or animal) relocation and expresses in a word whether the newcomers are accepted or repelled. The default position is that invasive alien species are undesirable newcomers, in particular if their dispersal is human-assisted; climate refugees are instead victims of anthropogenic change in nature. The victimhood implies that there must be a culpable party who owes something to the victims. Perhaps one acceptable, if not obligatory, way of repairing the moral relationship is to help the victim to survive, preferably in its current location or, if that is not possible, elsewhere. In other words, essential to the idea of assisted migration is the fact that conservables, such as species, populations or individuals, may not be able to survive without help provided by humans.

In general, it is important to ponder the nature of the responsibilities of humans whose actions in the form of global warming disturb ecosystem functioning and compel organisms to adapt or flee from their original habitats. The concept of refugee is a political one and presupposes the existence of a system of nation-states, territories and borders and the idea of citizenship; without the social reality as we today know it, refugeeship would not make much sense. In the borderless world, however, people could use their traditional "hunter-gatherer" adaptation strategy by migrating and taking important local flora and fauna with them. In this light, it does not seem a distant idea to apply the concept of refugee to non-human organisms, even though they are not persecuted for their convictions or ethnicity. It is equally interesting that the concept of citizenship seems to apply not only to humans but also for biological species, as their status changes after crossing national borders.

21.6   Concluding Remarks

In this chapter, we have examined conservation issues from the viewpoint of state sovereignty, and shown that problems may indeed in some cases arise. Biodiversity is a highly abstract idea embracing all biological variety above individual uniqueness on Earth. If humanity seeks conservation of this variety, the received wisdom says that collaboration between states is necessary. And when states collaborate and commit to the common guidelines for biodiversity conservation, they voluntarily narrow their scope for self-determination to some extent. The key aspect of sovereignty, however, remains. Most notably, if the states fail in implementation or have governments that break away from the successful policies of previous governments, they are subject to external critique in the form "naming and shaming". This has been particularly apparent in the fields of human rights and climate policies. In contrast to the human-rights framework, the possibility of military intervention for environmental reasons is virtually non-existent, although in some areas, poaching and wildlife trafficking have become a problem of a massive scale that are causing civil and park ranger casualties and, indeed, armed forces are being deployed from time to time. These difficulties in the implementation of conservation laws can be confronted partially by means of law enforcement and, therefore, the presence of civil society actions are vital for successful conservation. If so, a naturally arising idea is that nonstate actors should have opportunities to have an effect on international environmental legislation. As mentioned earlier, the topic is outside the scope of this chapter.

Although the compliance to international laws constrains the states' possibilities, sovereignty has still a key role in the actual drafting of conservation policies. States can decide on which populations, species and habitats they invest their conservation efforts. The states thus make priorisation decisions when such decisions need to be made. States can also open or close their gates to newcomers. States may even classify particular populations as species proper in cases where the majority of taxonomists recognise merely a subspecies. All in all, sovereignty is as noticeable in biodiversity conservation as in other areas of policymaking.

Acknowledgements We thank Minna Jokela, Elina Vaara and the editors of this volume for their insightful comments on the manuscript.

References

Andrews, R. N. L. (1999). *Managing the environment, managing ourselves. A history of American environmental policy*. New Haven: Yale University Press.

Beitz, C. (2009). *The Idea of Human Rights*. Oxford: Oxford University Press.

Birdlife. (2017). http://www.birdlife.org/worldwide/partnership/our-history. Accessed 30 May 2017.

Brunnée, J. (2008). Common areas, common heritage, and common concern. In D. Bodansky, J. Brunnée, & E. Hey (Eds.), *The Oxford handbook of international environmental law*. Oxford: Oxford University Press. https://doi.org/10.1093/oxfordhb/9780199552153.013.0023.

Endicott, T. (2010). The logic of freedom and power. In S. Besson & J. Tasioulas (Eds.), *The philosophy of international law* (pp. 245–259). Oxford: Oxford University Press.

Epstein, Y., López-Bao, J. V., & Chapron, G. (2016). A legal-ecological understanding of favorable conservation status for species in Europe. *Conservation Letters, 9*, 81–88.

European Commission. (2017). http://ec.europa.eu/environment/nature/legislation/habitatsdirective/index_en.htm. Accessed 12 June 2017.

Frank, D. J. (1997). Science, nature, and the globalization of the environment, 1870–1990. *Social Forces, 76*, 409–435.

Frankham, R., Briscoe, D. A., & Ballou, J. D. (2002). *Introduction to conservation genetics*. Cambridge: Cambridge University Press.

Gaston, K. J., & Spicer, J. I. (2004). *Biodiversity. An introduction* (2nd ed.). Oxford: Wiley.

Grove, R. H. (1995). *Green imperialism. Colonial expansion, tropical island Edens and the origins of environmentalism*, 1600–1860. Cambridge: Cambridge University Press.

Hällfors, M. H., Vaara, E. M., Hyvärinen, M., Oksanen, M., Schulman, L. E., Siipi, H., & Lehvävirta, S. (2014). Coming to terms with the concept of moving species threatened by climate change. A systematic review of the terminology and definitions. *PLoS One, 9*(7), e102979. https://doi.org/10.1371/journal.pone.0102979.

Hällfors, M. H., Aikio, S., Fronzek, S., Hellmann, J. J., Ryttäri, T., & Heikkinen, R. K. (2016). Assessing the need and potential of assisted migration using species distribution models. *Biological Conservation, 196*, 60–68. https://doi.org/10.1016/j.biocon.2016.01.031.

Hurme, E., Kurttila, M., Mönkkönen, M., Heinonen, T., & Pukkala, T. (2007). Maintenance of flying squirrel habitat and timber harvest: A site-specific spatial model in forest planning calculations. *Landscape Ecology, 22*, 243–256.

Kalliola, R. (1941). Luonnonsuojelusta ja sen tehtävistä. *Suomen Luonto, 1*, 15–24.

Kalliola, R. (1942). Foreword. *Suomen Luonto, 2*, 5–6.

Laine, T. (2015). *Suomen valkoselkätikkojen seurantaraportti 2010–2015. Linnut-vuosikirja, 12–19* (Summary: White-backed Woodpeckers in Finland: Monitoring report in 2010–2015).

Lyster, S. (1994). *International wildlife law. An analysis of international treaties concerned with the conservation of wildlife*. Cambridge: Cambridge University Press.

Mehtälä, J., & Vuorisalo, T. (2007). Conservation policy and the EU Habitats Directive: Favourable conservation status as a measure of conservation success. *European Environment, 17*, 363–375.

Nash, R. F. (2001). *Wilderness and the American mind* (4th ed.). New Haven: Yale University Press.

Nordqvist, O. (1899). Beitrag zur Kenntniss der isolierten Formen der Ringelrobbe (Phoca foetida Fabr.). *Acta Societatis pro Fauna et Flora Fennica XV*(7), 43 p. and appendices.

Oksanen, M., & Kumpula, A. (2018). Making sense of the human right to landscape. In M. Oksanen, A. Dodsworth, & S. O'Doherty (Eds.), *Environmental human rights. A political theory perspective* (pp. 105–123). London: Routledge.

Ostrom, E. (1998). Scales, polycentricity, and incentives: Designing complexity to govern complexity. In L. D. Guruswamy & J. A. McNeely (Eds.), *Protection of global biodiversity: Converging strategies* (pp. 149–167). Durham: Duke University Press.

Palmén, J. A. (1885). Internationelt ornitologiskt samarbete och Finlands andel deri. *Meddelanden af Societas pro Fauna et Flora Fennica 11*, 175–212.

Palmén, J. A. (1905). Luonnon muistomerkkien suojelemisesta. *Luonnon Ystävä, 9*, 145–153.

Palmgren, R. (1922). *Luonnonsuojelu ja kulttuuri 1*. Otava: Helsinki.

Parmesan, C., & Yohe, G. (2003). A globally coherent fingerprint of climate change impacts across natural systems. *Nature, 421*, 37–42.

Philpott, D. (2016). Sovereignty. In E. N. Zalta (Ed.), *The stanford encyclopedia of philosophy* (Summer 2016 Ed.). https://plato.stanford.edu/archives/sum2016/entries/sovereignty/. Accessed 31 Aug 2017.

Rusanen, P., Mikkola-Roos, M., & Ryttäri, T. (2011). Merimetsokannan kehitys ja vaikutuksia (Population development of cormorant and effects in Finland). Linnut-vuosikirja, pp. 116–123.

Sheail, J. (2010). *Nature's spectacle: The world's first national parks and protected places*. London: Earthscan.

Shue, H. (2014). Eroding sovereignty: The advance of principle. In *Climate justice. Vulnerability and protection* (pp. 141–161). Oxford: Oxford University Press.

Siipi, H., & Ahteensuu, M. (2016). Moral relevance of range and naturalness in assisted migration. *Environmental Values, 25*, 465–483. https://doi.org/10.3197/096327116X14661540759278.

Smith, I. A. (2016). *The intrinsic value of endangered species*. New York: Routledge.

Thomas, C. (2011a, October 29). Britain should welcome climate refugee species. *The New Scientist*. http://www.newscientist.com/article/mg21228365.600-britain-should-welcome-climate-refugee-species.html. Accessed 11 Sept 2018.

Thomas, C. D. (2011b). Translocation of species, climate change, and the end of trying to recreate past ecological communities. *Trends in Ecology and Evolution, 26*, 216–221.

Trouwborst, A. (2014). The habitats directive and climate change: Is the law climate proof? In C. Born, A. Cliquet, H. Schoukens, D. Misonne, & G. Van Hoorick (Eds.), *The habitats directive in its EU environmental law context. European nature's best hope?* (pp. 303–324). London: Routledge.

Turchetti, M. (2015). Jean Bodin. In E.N. Zalta (Ed.), *The Stanford encyclopedia of philosophy* (Spring 2015 Ed.). https://plato.stanford.edu/archives/spr2015/entries/bodin/. Accessed 31 Aug 2017.

UN General Assembly. (1962). Permanent Sovereignty over Natural Resources, General Assembly Resolution 1803 (XVII). Resolution 1803 (XVII). United Nations, 1962. http://legal.un.org/avl/ha/ga_1803/ga_1803.html. Accessed 30 May 2017.

Väisänen, R. A., Lammi, E., & Koskimies, P. (1998). *Muuttuva pesimälinnusto*. Helsinki: Otava.

Vuorisalo, T., & Laihonen, P. (2000). Biodiversity conservation in the north: History of habitat and species protection in Finland. *Annales Zoologici Fennici, 37*, 281–297.

Vuorisalo, T., Lehikoinen, E., & Lemmetyinen, R. (2015). The roots of Finnish avian ecology: From topographic studies to quantitative bird censuses. *Annales Zoologici Fennici, 52*, 313–324.

Wells, J. V., Robertson, B., Rosenberg, K. V., & Mehlman, D. W. (2010). Global versus local conservation focus of U.S. state agency endangered bird species lists. *PLoS ONE, 5*(1), e8608. https://doi.org/10.1371/journal.pone.0008608.

Wood, P. M. (1997). Biodiversity as the source of biological resources: A new look at biodiversity values. *Environmental Values, 6*, 251–268.

Permissions

The contributors of this book come from diverse backgrounds, making this book a truly international effort. We would like to thank all the contributing authors for lending their expertise to make the book truly unique. They have played a crucial role in the development of this book. Without their invaluable contributions this book wouldn't have been possible. They have made vital efforts to compile up to date information on the varied aspects of this subject to make this book a valuable addition to the collection of many professionals and students.

This book was conceptualized with the vision of imparting up-to-date and integrated information in this field. To ensure the same, a matchless editorial board was set up. Every individual on the board went through rigorous rounds of assessment to prove their worth. After which they invested a large part of their time researching and compiling the most relevant data for our readers.

The editorial board has been involved in producing this book since its inception. They have spent rigorous hours researching and exploring the diverse topics which have resulted in the successful publishing of this book. They have passed on their knowledge of decades through this book. To expedite this challenging task, the publisher supported the team at every step. A small team of assistant editors was also appointed to further simplify the editing procedure and attain best results for the readers.

Apart from the editorial board, the designing team has also invested a significant amount of their time in understanding the subject and creating the most relevant covers. They scrutinized every image to scout for the most suitable representation of the subject and create an appropriate cover for the book.

The publishing team has been an ardent support to the editorial, designing and production team. Their endless efforts to recruit the best for this project, has resulted in the accomplishment of this book. They are a veteran in the field of academics and their pool of knowledge is as vast as their experience in printing. Their expertise and guidance has proved useful at every step. Their uncompromising quality standards have made this book an exceptional effort. Their encouragement from time to time has been an inspiration for everyone.

The publisher and the editorial board hope that this book will prove to be a valuable piece of knowledge for students, practitioners and scholars across the globe.

Index

Printed in the USA
CPSIA information can be obtained
at www.ICGtesting.com
JSHW011400091023
49903JS00004B/38